D0078941

BRITAIN'S CHANGING ENVIRONMENT

FROM THE AIR

EDITED BY TIM BAYLISS-SMITH AND SUSAN OWENS

Britain's Changing Environment from the air

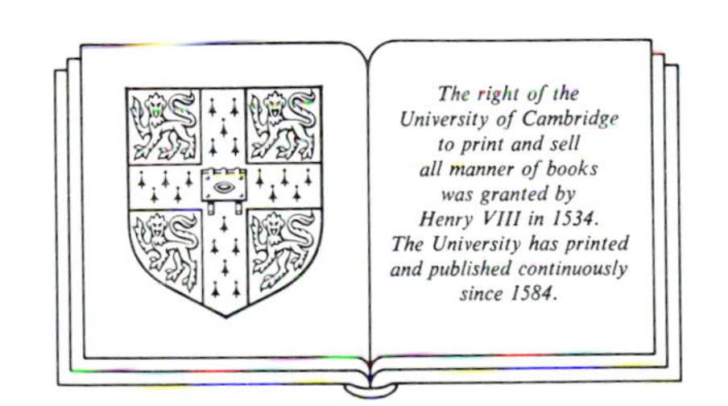

CAMBRIDGE UNIVERSITY PRESS
CAMBRIDGE
NEW YORK PORT CHESTER
MELBOURNE SYDNEY

Published by the Press Syndicate of the University of Cambridge
The Pitt Building, Trumpington Street, Cambridge CB2 1RP
40 West 20th Street, New York, NY 10011, USA
10 Stamford Road, Oakleigh, Melbourne 3166, Australia

First published 1990

Filmset and printed in Great Britain by BAS Printers Limited,
Over Wallop, Hampshire

British Library cataloguing in publication data

Britain's changing environment from the air.
1. Great Britain. Environment
I. Bayliss-Smith, Timothy P. II. Owens, Susan E.
333.70941

Library of Congress cataloguing in publication data

Britain's changing environment from the air
edited by Tim Bayliss-Smith and Susan Owens.
p cm
—Includes bibliographical references.
1. Aerial photography in geomorphology.
2. Geomorphology–Great Britain.
I. Bayliss-Smith, Tim II. Owens, Susan E.
GB400.42.A35875 1990
551.4′1′0941–dc20 90-200 CIP

ISBN 0 521 32712 1 hardback

Frontispiece: The small centre of Grasmere in Cumbria and the surrounding countryside receive the full impact of many thousands of visitors each year. (Cambridge University Committee for Aerial Photography, BYG8.)

Contents

Introduction: aerial photography, the hidden agenda

The conventional history of aerial photography starts with the first experiments from balloons, continues after 1909 when the first pictures were taken from an aeroplane, and then expands some 50 years later into the related fields of satellite imagery and remote sensing. In these accounts it is always the benefits to science, unforeseen but ultimately substantial, which receive all the attention. Cartography, geography and archaeology are singled out as beneficiaries, and this is not something we wish to dispute. Indeed, this volume on *Britain's Changing Environment* shows that for the study of landscape change the aerial photograph always provides fresh evidence. Very often it suggests, too, a new understanding of the processes at work. There is even a temptation to believe that what is seen from above has more reality than what faces one on the ground. It is a temptation that architects and planners in particular have learned to resist, for although the aerial photograph is one test of what their efforts have achieved, clearly the view from the air is not the real test. We do not live in aircraft, and the final test of the quality of an environment can only be made on the ground and by those who live there.[1]

In this Introduction we are not suggesting that the conventional account of aerial photography's benefits and pitfalls is wrong, but rather that it is incomplete. We therefore outline an alternative perspective on aerial photographs and the part that they play in our perception of the landscape and its resources. Alongside the positive contribution to science that everyone acknowledges, there is a more negative side to the story, and to explore this 'hidden agenda' will require us to look more closely at the military and propaganda uses of aerial photographs, and to examine their impact as a new form of knowledge. Rather than starting the clock in 1909 we see continuity between aerial photography and other techniques of landscape representation, some of which predate the origins of photography itself in the nineteenth century.

AERIAL PHOTOGRAPHY AND MILITARISM

The existence of a link between the production of images of landscape and the process of 'surveillance' seems undeniable. In relation to maps, Brian Harley has identified warfare, political propaganda, boundary making and the enforcement of law and order as the principal ways in which cartography as a form of knowledge has reinforced power relations.[2] Aerial photographs have been used in similar ways throughout this century, with military uses providing the primary motive for technological advance. The lenses and cameras that first made photography from aeroplanes feasible were pioneered in 1910–12 by the French army,[3] and in the First World War aerial surveillance rapidly became the main source

Aerial photography and the military: Pershore RAF station in Worcestershire, June 1941. The links between geography and militarism have been greatly strengthened by aerial photography, and more recently by satellite imagery. The view from the air transforms access to environmental information, and this knowledge reinforces power relations in a number of ways. In order to preserve their monopoly over this form of surveillance many national governments jealously defend their air space. As a result citizens' access to this source of knowledge is frustrated by state controls as well as financial cost.

of military intelligence. The potential of photographs for the accurate mapping of enemy terrain was also recognised, but only reached technical perfection during the Second World War.

Geography is one of the academic disciplines that has benefited most from this link with military surveillance. Just as, in the 1880s, the 'imperial importance of geographical knowledge' provided the initial means whereby professional geographers could establish for themselves a niche within the universities,[4] so, we would argue, the requirements of military intelligence in both world wars presented an opportunity for further academic expansion. Making maps and interpreting aerial photographs were, in fact, activities quite peripheral to the research and teaching interests of most geographers but, like remote sensing in recent times, these things provided in wartime a new means to power for the whole profession. As Balchin observed, 'we are accustomed to thinking of the subject as reaching towards peace and harmony among nations, but as the veil of secrecy is lifted we see that it also played a vital role in war'.[5] If this is so, then war has been equally important for the development of geography.

By the 1960s aerial photography had become only one of many techniques available for aerial surveillance. In 1969 General Westmoreland looked forward to the time when machines for remote sensing could replace American combat troops in Vietnam: 'I see an army built into and around an integrated area control system that exploits the advanced technology of communications, sensors, fire detection, and the required automatic data processing'. At the time this was a Star Wars fantasy: in Vietnam visual information was still of paramount importance, and the need for it justified a vast defoliation programme in areas of rain forest cover. However, by this time aerial photography was becoming absorbed into a much wider battery of surveillance techniques, as Congress was told:

> When one realizes that we can detect anything that perspires, moves, carries metal, makes a noise, or is hotter or colder than its surroundings, one begins to see the potential. This is the beginning of the instrumentation of the entire battlefield.[6]

Today there are particularly strong links between scientific research, remote sensing and the American Star Wars programme, and no doubt the same is true of its Soviet equivalent. Geographers and environmental scientists continue to be involved in this link, but generally in a rather passive role as the users of civilian satellite imagery too low in quality to have much value as military intelligence.[7] Much of their activity is essentially the inventory of environmental data, and their interpretation using information that is known, revealingly, as 'ground truth'. The basic questions that we would ask about this search for geographical insights through remote sensing are those that Michel Foucault would ask about any science:

> What is the history of this 'will to truth'?
> What are its effects?
> How is all this interwoven with relations of power?[8]

To answer these questions we must consider first the distinctive effect the aerial view has on environmental perception.

The bird's eye view, symbol of control over landscape: Bolsover Castle in Derbyshire, portrayed in the early eighteenth century as dominating its surrounding settlements and landholdings. In architecture there is clear continuity between this elegant and detached view of the relationship between property and resources, and the aerial perspective that is achieved by photography. In both cases dirt, poverty and oppression are rendered invisible and the landscape is reduced to a geometrical pattern.

AERIAL PHOTOGRAPHY AS PROPAGANDA

Even if we convince ourselves that our research efforts using aerial photographs are not implicated in the power relations of imperialism or repression, we still need to be alert to some intrinsic properties of photographs as information sources. The view from the air differs from the ground view not merely in its visual perspective but also in the sense of detachment that it provides. Kenneth Hudson remarked on this in his introduction to an earlier volume in the Cambridge Air Surveys, noting that the view from the air always seems to dehumanise the landscape, particularly the urban landscape.[9] These feelings have been noted by almost everyone involved in flying:

> Flying alone! Nothing gives such a sense of mastery over mechanism, mastery indeed over space, time and life itself, as this . . . From ten thousand feet how insignificant the work of men's hands appear! How everything they do seems to disfigure the face of the earth; but when they have done their worst, what a lot of it is left![10]

This pilot recalled how the Western Front in 1917 appeared merely as 'lines, drawn anyhow as a child might scrawl with a double pencil'. From the air the landscape was reduced to 'an ordnance map, dim green and yellow', while the battlefield itself seemed robbed of all significance.

The subtle effects of this perception on the conduct of warfare have often been noted. Another pilot recalls night bombing in 1943:

> I would try to tell myself that this was a city, a place inhabited by beings such as ourselves, a place with the familiar sights of civilisation. But this thought would carry little conviction. A German city was always this, this hellish picture of flame, gunfire and searchlights, an unreal picture because we could not hear it or feel it breathe.[11]

Even in daylight and with no ground retaliation, it requires some effort when flying 6km above the ground to imagine what really is 'ground truth'. During the American war against Vietnam and Laos 'every civilian on the ground was assumed to be enemy by the pilots by nature of living in a free-fire zone'.[12] Reports by Western observers on the effects of aerial bombing were restricted to areas under American control, and the pilots continued to carry out their orders.

Another consequence of the detached and dehumanised character of the aerial view is that it can be an effective way of conveying an abstract or even distorted view of reality. Aerial photographs, like maps,[13] can thus become subtle tools for propaganda. The device can be traced back to the seventeenth century, when artists were commissioned by landowners to show the 'bird's eye view' of a property and its surroundings.[14] These aerial pictures of country estates and, at a later date, factories were not merely instruments of power, serving to demonstrate the scale of architectural achievement and territorial control. Their content also conveyed a subliminal message, the bird's eye view providing an attractive, clean and acceptable image of property ownership, agricultural enclosure or the factory system. From high above, the poverty, dirt and social injustice which accompanied the creation of these landscapes are simply not visible.

Architects have also learned to use the bird's eye view, but as a more explicit device for persuasion. This is particularly well demonstrated by those gigantic urban building projects that so fascinated the profession in the 1950s and 1960s, the so-called megastructures. Too vast and complex to make much sense unless viewed from above, these schemes almost always required also an idealised and detached viewpoint in order to make them appear acceptable. In Britain those that have been built have been notable failures in human terms: Cumbernauld Town Centre and Thamesmead New Town are prime examples. Those who promoted such schemes would superimpose architectural models of the proposed buildings on an actual aerial photograph or map of the existing landscape, and the entire project was then re-photographed from above. In the words of Reyner Banham, historian of modern architecture, this technique for display reduced a scheme to 'a giant diagram imposing an architectural, cultural, social and possibly political order on a vast stretch of countryside'.[15] He argues that for a while this combination of implied advantages proved almost irresistible to politicians, planners and developers, and that the detached way in which megastructures were presented to decision-makers enhanced their acceptability. Although Prince Charles is strongly opposed to the whole modern movement in architecture, it is interesting to note that in his recent book he also uses numerous aerial views, some idealised, some photographic, in order to project his alternative vision.[16]

Aerial photographs usually tell fewer lies about the actual appearance of things than the bird's eye view, but they do tend to make almost any pattern on the ground look pretty. The combination of their authenticity, their claim to objectivity and their intrinsic aura of detachment creates new opportunites for disinformation. One can see the roots of this tendency in the very first landscape photographs. An early example is the book *Illustrations of China and its People* by John Thomson (1873), a pioneer of 'realist' photography, and based on over 200 photographs taken during five years of travel. Thomson was a Fellow of the Royal Geographical Society, and the ostensible purpose of his book was simply to provide geographical information, the pictures offering 'the nearest approach that can be made towards placing the reader actually before the scene'.[17] In fact, as Ian Jeffrey has pointed out, Thomson was surprisingly uninterested in mere landscape:

> His book is no disinterested survey. It is rather more of a prospectus, designed to be of use to traders and to settlers. Travelling along the Yangtze he was continually on the alert for steamboat routes and settlement sites . . . He was agitated by the sight of such untapped human and mineral resources and produced what is, in effect, a colonizer's handbook.[18]

Illustrations of China is perhaps the first major example of the camera being used to convey a particular version of geographical 'ground truth'. Aerial photographs became available for the same purpose in the 1920s, and are used today in more or less explicit ways – for instance, by the oil industry promoting its 'green' image through advertisements that show how invisible underground pipelines are, even when viewed from the air. Other examples abound: in 1988 three picture-books were published showing China, South Africa and New York, each viewed exclusively from the air, and in different ways conveying a less than complete picture of ground truth. *The Guardian* newspaper commented:

> Aerial photography has become the supreme tourist board way to sell the idea of a country. No dirt, no poverty, no people to be seen; no waste, tensions or ugliness – Earth a series of abstract colours, shapes and shadows of man's imprint. The higher the Cessna or balloon from which the photographs are taken, the more removed and unreal the planet becomes.

In this book the photographs are in monochrome rather than colour, and their ideological purpose is quite different, so we hope that reviewers will spare us from this particular criticism!

Nevertheless we must acknowledge that photography is not a neutral means of representation. John Tagg argues that its meaning can only be understood within a particular historical context:

> Photography as such has no identity. Its status as a technology varies with the power relations which invest it. Its nature as a practice depends on the institutions and agents which define it and set it to work.[19]

We therefore suggest that any claim that the use of aerial photography achieves a neutrality of vision is patently false. In every collection of photographs, this one included, there has been a careful selection carried out in relation to a particular didactic purpose. The pictures illustrate particular points that the authors wish to make. If different points were intended, even if perhaps the opposite point of view had been taken, then undoubtedly different photographs could have been found, providing alternative visual images that perhaps would have been equally convincing.

AERIAL PHOTOGRAPHY AS KNOWLEDGE

Aerial photography has had an impact in a wide range of fields. The scale of this impact makes more sense if we consider the status of photography as a new technology for the production of knowledge. Michel Foucault has argued that the production of any new kind of knowledge will inevitably release new effects of power, just as the new ways in which power is then exercised will themselves generate further knowledge.[20] The development of photography in the nineteenth century is a good example of this mutual reinforcement of technology and power.

In his book *The Burden of Representation*, John Tagg has shown how, from the 1870s onwards, photographs provided state institutions with a new instrument of evidence concerning the identity of individuals. Photography provided the means whereby new techniques of surveillance and record keeping could be developed. Tagg argues that the whole network of disciplinary institutions in industrialised societies – the police, prisons, asylums, hospitals, even schools and factories – required an enormous expansion in social administration, itself scarcely imaginable without new technology, a key component of which was the photograph.[21]

As we have shown, the enormous political and military implications of aerial surveillance mean that the history of aerial photography is basically a history of knowledge as power, or what Tagg calls 'a rendezvous between a novel form of the state and a new and developing technology of knowledge'. The process was actually foreseen by the pioneers of photography in the early nineteenth century. No one doubted at the time of its invention that photography was going to cause a revolution in environmental perception. On first seeing a daguerrotype image Alexander von Humboldt realised at once what the invention signified. 'Daguerre is my Chimborazo', he wrote in an enthusiastic letter to Fox Talbot, the English pioneer in photographic techniques.[22] This reference to the great tropical mountain, which in Humboldt's *Cosmos* serves as the complete illustration of all the world's climatic regions and vegetation belts, reveals the scientist's recognition of the immense advance in the ease and precision of image portrayal that photography represented. As Lady Eastlake wrote,

> For everything for which Art, so-called, has hitherto been the means but not the end, photography is the allotted agent – for all that requires mere manual correctness and mere manual slavery . . . She is made for the present age, in which the desire for art resides in a small minority, but the craving, or rather necessity for cheap, prompt and correct facts in the public at large. Photography is the purveyor of such knowledge to the world at large.[23]

The impact of these 'cheap, prompt and correct facts' about the world's landscapes that had suddenly become available was virtually immediate. As early as 1840 *Excursions daguerriennes* was on sale in Paris, a collection of engravings based on 1,200 daguerrotypes of landscapes from Niagara to Egypt and Moscow.[24] None of this was aerial photography in the strict sense, but one cannot look at these early daguerrotype views, many of them from high buildings overlooking the rooftops of Paris, Edinburgh, Jerusalem and Beirut, without experiencing the same insights that are intrinsic to any elevated viewpoint. As the century progressed, satisfying the demand for landscape photographs developed into an industry, with picture-books, postcards and, after 1860, stereoscopic views growing in popularity. In a very real sense the study of geography itself became possible, as the mere description of landscape had become a mundane task and the much more challenging task of interpretation became feasible.

Today, more than a century after geography's establishment as an academic discipline, and 150 years after *Excursions daguerriennes* appeared in Paris, the search for pattern and process in the landscape continues. This present volume is partly motivated , like all its predecessors, by a simple fascination at the images of landscape. However, in the pages that follow there is also conveyed a growing sense of urgency and unease about what the photographs reveal. In every chapter a concern is expressed for what is happening to the landscape of Britain, and what processes underlie those changes so strikingly demonstrated from the air. In Britain, as elsewhere, there is inescapable evidence of increasingly rapid and perhaps irreversible changes in environmental systems, and this realisation is at last becoming widespread. If aerial photography can be the vehicle for our own concern about the environment's degradation to be communicated to an even wider audience, then we shall be well satisfied.

1 The changing uplands

The mountains, moorlands and hills of Britain are diverse both in terms of scenery and of ecology. Altitude brings limits in ecological productivity, and these areas tend to have distinctive kinds of natural vegetation, notably heath made up of dwarf shrubs like heather (*Calluna vulgaris*) and grasslands. There are typically few naturally regenerating trees above about 450m altitude. Although they are frequently portrayed as both unchanging and unused, upland environments have in fact been extensively influenced by people over a long period, and have a series of important economic roles to play in modern Britain (photo 1). Most are grazed by domestic stock or game animals. Recreation, water supply, mineral extraction, military training and coniferous afforestation are other important, although controversial, uses of upland environments.[1] Hill and moorland scenery has earned protection through National Parks and Areas of Outstanding Natural Beauty (AONBs) in England and Wales, and in National Scenic Areas in Scotland. The importance of the various upland vegetation communities is recognised in the designation of areas of land as National Nature Reserves (NNRs) and Sites of Special Scientific Interest (SSSIs).[2]

The very diversity of the British upland areas makes any definition of their limits an arbitrary one. It is hard to draw out reliable generalisations about such varied areas. Thus, although each is undeniably 'upland' in some sense, there is a vast difference between the relatively mild conditions on Exmoor at 450m and the tundra-like plateau of the Cairngorms at 1,200m. Words such as 'hill', 'moor' or 'mountain' have a range of both general and specific meanings in different contexts, and are inadequate for topographic description. The word 'moor' for example is defined by the OED as a 'tract of open waste ground especially if covered with heather: tract of ground preserved for shooting'. There is no necessary correlation with altitude.

In a study of upland vegetation change in 12 upland parishes in England and Wales,[3] The Institute of Terrestrial Ecology defined three main categories of land: upland margin between about 120m and 245m, upland between about 245m and 335m and hill land between about 430m and 610m. There are problems even with this careful definition. Certainly the 245m contour in Britain corresponds fairly closely to the boundary of the Less Favoured Areas (LFAs) in England and Wales (similar to the old category of 'hill' land, where production was constrained by environmental conditions). It was also used to define the limits of the Countryside Commission's upland policy report in 1984.[4] However, such a definition on the basis of altitude makes less sense when Scotland is included, because altitude is not the only important factor. On the Parphe in the far north-west of Scotland, vegetation communities occur at about 300m or below which are typical of 900–1,200m mountains elsewhere.[5] The 245m contour also fails to

1 *opposite* Ben Nevis and Fort William. At 1344m Ben Nevis is the highest point in the British Isles, rising from sea level at Fort William. Ski development is now taking place on nearby Aonach Mor.

2 Loch Avon and the Cairngorm Plateau, looking up Loch Avon towards the slopes of Ben Macdui which lies out of the picture on the left above the steep crags which line the head of the loch. There are still snow patches on the plateau, a reminder of the harsh climate. The vegetation of Cairngorm is roughly zoned with altitude, and on the plateau there is a spare moss heath with close affinities to vegetation in the Arctic. Loch Avon and part of the plateau are now owned by the Royal Society for the Protection of Birds.

take account of important differences in the vegetation of montane and submontane zones. A flexible definition is therefore required, recognising a lower boundary of 245m but also taking into account the gradation in environmental conditions above that, and the nature of the vegetation and the land use (photo 2).

A broad definition of this kind effectively divides Britain in two. Upland Britain lies to the north and west of a line between the Rivers Severn and Trent, the lowlands to the south and east. Much of Scotland and Wales is upland in character, and a rather smaller part of England. In Scotland the many mountain blocks of the Highlands, the northwest coast, the Inner Hebrides, Outer Isles,

3 Hirnant in the Berwyn in North Wales. The complex mosaic of valley and hill land is clearly visible. Small hedged fields lead up to the open moorland. This is grazing country, and few of the fields are cultivated. Farms are small and isolated. The moorland is important for conservation, and one of the last strongholds of the merlin. However, it is also suitable for afforestation, and lower areas can be turned into improved pasture. In the early 1980s forestry, farming and conservation came into conflict in the Berwyn, farming moving up onto the open moor, and forestry moving down. Both threatened open moorland. There was strong opposition from the local Berwyn Society to the proposed designation of large areas as Sites of Special Scientific Interest. The Nature Conservancy Council appointed a consultant to come up with a compromise land use plan which attempted to partition areas between the three interests. This eventually won support from local farmers, but remained controversial for conservationists because of the relatively low priority given to conservation.

Orkney and Shetland and the Southern Uplands and Cheviots are all included. In Wales there are large areas outside the well-known areas of Snowdonia, the Black Mountains and the Brecon Beacons, such as the Berwyn (photo 3) and the extensive Cambrian Mountains. In England upland areas include the Northumberland Fells and North Pennines, the Lake District, the Yorkshire Moors and Dales, the Peak District, Dartmoor, Exmoor and Bodmin Moor. It is probably these English hills, some discrete blocks of which have National Park status, that are familiar to the largest number of people in Britain. In fact they represent only a small part of the total upland area of Britain, and a number of the examples in this chapter are drawn from what may be less widely known areas.

UPLAND ENVIRONMENTS

4 *opposite* The ridge of Beinn Eighe (foreground) and Liathach (behind) in Wester Ross. Behind Liathach lies the head of Loch Torridon.

Environmental conditions in these upland areas vary considerably, but certain features are common. Geologically, upland areas tend to be underlain by hard slowly-weathered rocks. Often these are igneous like the granite of Dartmoor, or metamorphic like the slates of parts of Snowdonia or the Lake District, the gneiss of Sutherland or the quarzite capping on the Torridonian peaks of Beinn Eighe and Liathach (photo 4). In general these rocks provide poor soils for vegetation growth, although there are glorious exceptions where lime-rich strata support rich arctic and alpine floras.[6] More usually, mountain soils are shallow, unstable

and immature, eroding rapidly and intensely leached. They tend to be acidic, particularly deficient in calcium and phosphorus.

Climatic conditions in the uplands are also harsh, with low temperatures, high precipitation and high winds. Mean temperature falls 1°C with every 100m of altitude, while windspeed, the amount of cloud cover, the total precipitation and the number of wet days, all increase with altitude. The extent and duration of snow cover also rise. Altitudinal variations in environmental conditions are combined with more general trends across the country, with a rise in the amount and uniformity of precipitation and the incidence of gales towards the west, and falling temperatures to the north.

It is not surprising that these variations in environmental conditions are closely reflected in the vegetation, both in the floristic composition of plant communities and in their productivity. Plants differ in their resistance to cold, but for each species there is a lower temperature limit below which it is unable to remain active, and a further limit below which it dies. Freezing conditions cause the formation of ice in the tissues, and hence the physical rupture of cells and the loss of water essential to vital functions. Mountain plants often show adaptations to cold in the form of small size, cushion, mat or tussock growth forms. Some are annuals, passing the long winter in the form of seeds, others are perennial, with woody tissue resistant to frost, and mechanisms of frost-hardening protecting sensitive growing points. Trees at altitude are frequently dwarfed and deformed in growth form, and tree lines are determined by the adaptation of woody species to frost, although mechanical stress and desiccation are also important. In much of Britain the natural tree line is probably around 600m, falling to 300m in Scotland, although as Pears found in the Cairngorms, grazing and burning have created an artificially lowered tree line almost everywhere.[7]

Human impacts are significant in almost every part of the British uplands. Even the Cairngorm plateau is grazed, by sheep and introduced reindeer (photo 5). Only remote mountain cliffs escape grazing, and upland vegetation is generally referred to as 'semi-natural'. Despite this human influence, vegetation patterns reflect the natural environment quite closely. Classic work on the Cairngorms by Watt and Jones shows a shift from heather to bilberry and crowberry dwarf heaths in the lower regions eventually to moss heaths in the alpine zone, although there is a great deal of variation, including extensive bog communities in wet locations.

Not only do species tend to differ at different altitudes, but their luxuriance and size tend to decline. Studies in connection with the International Biological Programme, for example, showed clearly that the rate at which plants comprising natural vegetation grew, and hence their total growth in one season, declined with altitude.[8] One reason for this is the lower average temperatures, but of greater importance is the shorter growing season, in other words the fact that temperatures permit plant growth for a shorter period at altitude.

Work on the Moor House National Nature Reserve, on the highest point on the Pennines, showed that the productivity of heather in a bog environment declined with altitude and at 850m was only 20 per cent of its productivity at 550m.[9] Other studies in the Cairngorms at altitudes over 850m showed that production varied with the degree of exposure, and that *Calluna* plants taken from

5 View down Glen Derry towards the caledonian pine woodlands of Derry Lodge, Deeside, Cairngorms.

high, exposed positions were adapted to survive there, and could respire faster at lower temperatures and over a wider temperature range.[10]

The low productivity of upland environments has long been recognised. There have been periods in prehistory when more benign climatic conditions prevailed in the hills, and crop cultivation was possible. The wealth of archaeological remains on the now bleak and inhospitable peaty soils of Dartmoor, such as the Bronze Age village of Grimspound in the catchment of the East Dart, are evidence of this. There have been inroads of cultivation into the hills on many subsequent occasions, and the abandonment of reclaimed land, in response to improved conditions or economic circumstances. In general, however, the uplands have been relegated to the economic margin, and used primarily for extensive forms of land use, notably grazing. In the past, high pastures were grazed in the summer months by cattle under a transhumance system. In the nineteenth century this pattern gave way to one of large sheep farms and, rather later, the shooting estate, and in Scotland the deer forest. In this century the deer and grouse have given a little ground to the walker and climber, but sheep have remained. Hill farming has been subsidised by the government since the Second World War, but overall the uplands have been areas of economic decline and out-migration.

Forestry has expanded as hill farming has declined, in both the public and private sectors. It now competes with hill farming and conservation interests for plantable land. Afforestation has had a major effect on upland landscapes, particularly where extensive continuous coniferous forests have been developed. This competition presents what is probably one of the largest single challenges to conservation in Britain today, and it is the product of confused policies over a long period. However, there are other competing interests in the uplands. Mineral extraction, military training and water supply have all been important in moulding present landscapes, and all remain controversial. Measures for the conservation of landscape and wildlife, and the planning of land use in the uplands, have developed greatly in the last 40 years. Today, questions about the future of the uplands are of considerable interest and considerable importance.

UPLAND VEGETATION

Despite first appearances, upland landscapes are far from unchanging. There have been major natural changes in vegetation communities over the last 10,000 or so years, revealed through the analysis of pollen and plant macrofossils in peat

6 Part of the Rannoch Moor. Rannoch Moor lies north of Tyndrum and east of Glen Coe. The photograph shows clearly the diverse vegetation of the surface of the peatlands, with patterns of pools and hummocks built up on the peat. Pollen analysis shows that after the retreat of the ice 10,000 years ago, this area was colonised by birch followed by a mixed woodland of pine, birch, hazel and alder, before wetter conditions led to the development of blanket bog. The railway to Fort William can be seen running from right to left across the upper third of the picture. The peat offered a considerable obstacle to the construction of a sound track bed for this line, and the line remains remote. In the time since this photograph was taken, the area beyond the line has been planted up with conifers, destroying the palaeoenvironmental archive and the vegetation community of the bog. Part of Rannoch Moor is a National Nature Reserve.

deposits.[11] On Rannoch Moor, for example, it would appear that the ice disappeared around 10,000 years ago, and a dwarf shrub heath of crowberry and juniper developed (photo 6). By 9,800 years ago, birch woodland was established, followed by a forest of birch, hazel, scots pine and alder.[12] Written records provide evidence of more recent changes within the historical period. Information on woodlands is particularly good.[13]

Traditional views have suggested a catastrophic loss of upland forests in the eighteenth and nineteenth centuries, partly on the basis of contemporary complaints about activities such as iron smelting. The evidence now is rather for a more sustained use, especially of the more productive woods at lower altitude. Coppicing of oakwoods for charcoal manufacture began in Scotland in the sixteenth century, for example for iron foundries in Argyll and in the oakwoods around Lock Maree in Wester Ross. Coppicing spread in the seventeenth century,[14] at which time exploitation of the native scots pine woodlands also began. By the late eighteenth and nineteenth centuries, coppicing to produce charcoal and oak bark for tanning was extensive in the Highlands and in many other parts of upland Britain.

Fragments of forest or woodland remain in upland areas, and are of considerable ecological interest and conservation importance. The sessile oak *Quercus petraea* is typical of base-poor soils of upland valleys of Wales and England as well as parts of Scotland. The pedunculate oak *Quercus robur* also occurs in the uplands, most notably on Dartmoor in Wistman's Wood and Black Tor Copse. Birch (*Betula pubescens*) is common on acid upland soils, sometimes (although by no means always) occurring above oak (photo 7). Fragments of birch woodland survive on the Inverpolly National Nature reserve in Wester Ross, and there is a particularly fine example on Morrone near Braemar rising to 610m.[15] Birch often

7 Craigellachie National Nature Reserve, near Aviemore in Speyside. The reserve is primarily declared as a birch woodland, and the photograph shows the open structure of the woodland and the continuity of vegetation community up onto the moorland above. The wood contains both the 'silver' birches, *Betula pubescens* and *Betula pendula* as well as trees such as Rowan (*Sorbus acuparia*). The understory is variable, including bracken (*Pteridium aquilinum*), wood rush (*Luzula sylvatica*) and heaths of ling (*Calluna vulgaris*) and *Vaccinium* species. There is also a range of wetland environments visible within the wood. In the background the hard edge of a conifer plantation can be seen clearly, the contrast with the diverse and light birch woodland in the foreground being very clear.

8 *opposite* Loch Rannoch and the Black Wood of Rannoch. The Black Wood is one of the surviving fragments of caledonian pine forest which once covered large areas of the central highlands. This is dominated by one of the three native conifers in Britain, the scots pine, *Pinus sylvatica* (the others are gorse and juniper). Native scots pine now exists only in small patches, and many of these have been affected by cutting, planting and deer grazing in the last 200 years. Analysis of resins suggests, however, that some stands are more or less pure, and certainly the open woodland structure is characteristic, and quite different from that of the dank closed conifer plantations with which caledonian pine forest is often replaced. The photograph shows a mixture of indigenous caledonian pine forest and other subsequent plantations. A forestry road makes a scar across the foreground. Much of this area has now been converted to plantation. In the distance lies the open extent of Rannoch Moor, itself now partly planted up with exotic conifers (see photograph 6).

occurs with rowan (*Sorbus acuparia*) and hazel (*Corylus avellana*), with alder (*Alnus glutinosa*) on wetter sites. Ash can be important in upland woods on more base-rich soils, for example the Helbeck and Swindale Woods of the Pennines (rising to 360m), and the Rassal Ashwood of Wester Ross near sea level.

Scotland also has important areas of native scots pine woodland (*Pinus sylvestris*) surviving.[16] Scots pine is a planted tree in England and Wales, but native in Scotland. Although some replanting has been done, and continental genetic stock has been introduced, some Scottish pinewoods are relatively unaltered by man. Documentary evidence, soil structure, the age-structure of the trees and the diversity of associated plant and animal species suggest continuity with the wildwood of the past. Only some 11,000ha survive, of which only 1,300ha is high forest. There are good examples in the Spey and Dee Valleys, in the Black Wood of Rannoch, and on the shores and islands of Loch Maree, although some felling continues (photo 8). The small size and fragmentation of surviving stands, combined with heavy grazing pressure, bring fears for their long-term viability.[17] The Nature Conservancy Council is now having some success in promoting regeneration by excluding deer from fenced enclosures, for example in the Cairngorms and the Beinn Eighe National Nature reserve in Wester Ross. In 1987 the Royal Society for the Protection of Birds acquired a substantial area of the Abernethy Forest in Speyside as a reserve.

Forest clearance followed by grazing has led in many upland areas to an artificially lowered tree line and the fragmentation of the zone of scrub species such as hazel (*Corylus avellana*), juniper (*Juniperus communis*) and willows. In places in Scotland fragments of typical arctic scrub species, the downy willow (*Salix lapponum*) and woolly willow (*Salix lanata*), still occur. But largely replacing these species are extensive areas of dwarf shrub heaths, many of whose species would have survived as the understory of forest. Heather or ling is the most familiar of these species. Heather moors occur widely, most notably perhaps on Dartmoor and parts of Exmoor, the North York Moors and the drier mountains of the eastern Highlands. Where sheep grazing is extensive, heather moors tend to be replaced by grassland, typically a mix of bents such as *Agrostis tenuis* and *Agrostis canina* and fescues such as *Festuca ovina*. Heavier grazing, particularly on wetter soils, favours grasses such as mat grass (*Nardus stricta*), purple moor grass (*Molinia caerulea*) and the heath rush (*Juncus squarrosus*). There are extensive grass moors in the hills of Galloway, Northumberland and central Wales. In places *Calluna* is replaced by other species in the same general dwarf shrub vegetation. Thus gorse (*Ulex gallii*) is important in southwest England, crowberry (*Empetrum nigrum*) and cowberry (*Vaccinium vitis-idaea*) in the Peak District and bearberry (*Arctostaphylos uva-ursi*) in the eastern Highlands. Locally, bracken can be an important component of vegetation.

At higher altitudes distinctive vegetation occurs in areas where snow lies, bilberry (*Vaccinium myrtillus*), *Nardus* and certain mosses such as *Polytricum norvegicum* being common in places. On higher mountains which rise above the dwarf heath zone, various moss heaths and fellfield communities occur. Cushion plants such as moss campion (*Silene acaulis*), the downy willow (*Salix lapponum*), mountain avens (*Dryas octopetala*), various saxifrages and the stiff sedge (*Carex bigelowii*) are common in this zone.

9 *top* Cow Green Reservoir, Upper Teesdale. The bright concrete of the dam stands out clearly in this photograph, as does the access road and the bare shoreline eroded by wave action. In the background the peak of Cross fell, at 2930m the highest point in the Pennines, can be seen. Cow Green was built by the Tyne-Tees Water Company to supply water to ICI on Teesside, and finished in 1970. It was one of several sites investigated in the 1960s, and unhappily flooded part of Widdybank Fell, a site of Special Scientific Interest due to its unique assemblage of arctic-alpine plants. These include the spring gentian, mountain avens and spring sandwort. The vegetation community developed on a partly metamorphosed limestone, called sugar limestone, and the plants may have survived *in situ* since the end of the last glacial period (about 10,000 years ago) because of the friable nature of this rock as well as the bleakness of the Pennines. Construction of the reservoir was highly controversial, and became the subject of a Parliamentary Bill. Despite the undoubted importance of the site, construction went ahead. Part of the arctic-alpine vegetation was lost, and more has been lost subsequently by bank erosion. On the positive side some research by the Nature Conservancy Council was financed by the developers, and more is known of what existed, and what is left. Research has also been carried out by the Freshwater Biological Association on the impact of the dam on the flora and fauna of the river itself below the waterfall of Cauldron Snout just below the dam. Ironically, the subsequent construction of the much larger reservoir at Kielder has made the provision at Cow Green superfluous to water supply needs in the Northeast, so the loss of part of the SSSI has been in vain.

In parallel with these communities of drier moorland and mountain areas, distinct bog vegetation communities develop in wet conditions in association with the formation of peat. Upland peatland terminology is complex,[18] but a basic distinction can be drawn between peat formed as a result of impeded drainage, such as valley mires, and blanket peat which forms independently of topography and is sustained by precipitation. The Caithness flows are probably the finest examples of blanket peats of this kind, and until their very recent reduction by forestry they represented the largest area of virtually unmodified vegetation in Britain.[19] Other extensive areas of blanket peat exist in the Pennines and on Dartmoor. Blanket peatlands can grade into wet heaths, and are typically dominated by the deer sedge (*Trichopherum caespitosum*), heather (*Calluna vulgaris*), the purple moor grass (*Molinia caerulea*), heath rush (*Juncus squarrosus*) and bog cotton (*Eriophorum vaginatum*). The cross-leaved heath (*Erica tetralix*), bog asphodel (*Narthecium ossifragum*) and bog myrtle (*Myrica gale*) occur widely. Valley mires such as Kirkonnell Flowe or Silver Flowe in Galloway have similar vegetation, but are notable for the development of fine hummock and hillock topography, which is based on the continuous succession of different species of sphagnum moss.

Species diversity on mountains tends to be greatest where lines of downslope water movement concentrate nutrients in 'flushes', and where grazing pressure is low, for example on inaccesible cliff ledges. The montane flora of Britain is best developed where such base-rich rocks occur at altitude. One such location is in Upper Teesdale, where a unique assemblage of arctic and alpine species is concentrated in an area of sugar limestone[20] (photo 9).

Many species, such as dwarf birch (*Betula nana*), spring gentian (*Gentiana verna*) and birds-eye primrose (*Primula farinosa*) are extremely restricted in their distribution in Britain, and the combination of species whose normal distribution is restricted to arctic or alpine areas with others from southern Europe is of enormous interest. The Dalradian Schists of Ben Lawers in Perthshire support a similar abundance of rare species, and make this another mecca for botanists (photo 10). There are important areas elsewhere, for example on the Durness Limestone of Sutherland.[21]

THE ECOLOGY OF UPLAND MANAGEMENT

Ecologically, most upland areas are held in a stage of arrested succession by grazing and burning. Ecological change can result from either an increase or a decrease in the intensity of either activity. Heather provides rather poor food for sheep except in winter when annual grasses are exhausted, but is essential in the diet of the red grouse (*Lagopus lagopus scoticus*), which has the distinction of being Britain's only endemic bird. The nutritional value of heather declines with age, and in order to maximise grouse numbers on shooting moors heather is regularly burned to promote regrowth of young shoots. Grouse also require mature heather for nesting, and a diverse age structure in the heather is optimal, with burned patches of about 1ha burned perhaps every 10 years.[22] Intensively managed grouse moors tend to have a distinctive striped appearance created by the scars of successive burns (photo 11). Excessively hot fires (caused by burning when there is too much wood in the heather, or when it is too dry) or over-frequent

10 *opposite, below* Beinn A'Ghlo. Hill in central Perthshire important for its arctic-alpine plant communities.

burning leads to the replacement of heather with other species, including gorse and bracken, and some grasses.

Grasses can also invade heather if it is intensively grazed. On dry soils a bent/fescue grassland may develop which is likely to be more productive for sheep, but on wetter soils the mat grass (*Nardus stricta*) and rushes (*Juncus* spp.) come to dominate, and these are of little nutritive value. Grazed grasslands yield no grouse, and have a greatly restricted diversity of flowering plants and hence very limited nature conservation interest. Further increases in grazing intensity also lead to the gradual eradication of more palatable grasses such as *Festuca* and *Agrostis* species and the rise of less nutritious grasses and rushes. Overgrazing can lead quite rapidly to the decline of the value of upland pastures.

The effects of grazing on upland vegetation can be quite dramatic. The area of heather moorland in the Peak District declined by 36 per cent between 1913 and 1980, and this has been linked with a trebling in sheep numbers between 1930 and 1976.[23] Experiments in which sheep are excluded from sample plots show that after 10–15 years the more succulent grasses reappear, followed by the dwarf shrubs. Studies of a grazed blanket bog on the Moor House National Nature Reserve in the Pennines, for example, showed a recovery of heather after 15 years,[24] while on the sugar limestone of Upper Teesdale species diversity increased and species such as mountain avens (*Dryas octopetala*) and the hoary rock rose (*Helianthemum canum*) reappeared after 12 years.[25] Where pastures have been enclosed and improved, former vegetation can take considerably longer to recover. Studies by the Institute of Terrestrial Ecology examined the extent to which improved pastures abandoned at different periods in the past had reverted to dwarf shrub heath. Only 20 per cent of the area abandoned in the period 1941–78 had reverted by 1980, although a further 40 per cent was on the way there as a grassy heath mixture. Of pastures abandoned before 1850 only 35 per cent had fully reverted to shrub heath vegetation.[26]

This evidence highlights the sensitivity of upland vegetation communities to changes in management and land use. Both their economic value and their conservation interest can quickly be affected by changes in management regime, such as stocking density or frequency of burning. These, in turn, can be the result of social or economic change in upland areas, for example, the enclosure and improvement of open moorland or the abandonment of improved pastures and the end of intensive management. Major changes in land use can be made very rapidly, but their ecological effects remain for decades and even centuries, even if the land reverts to its previous use. Considerable attention has therefore been directed in recent years to the measurement of upland land use change. Surprisingly little detailed information in fact exists on changes prior to the 1960s and 1970s, when such studies began.

11 *opposite* Patches of burned heather on a managed grouse moorland. This area lies a few miles north of Balmoral on Deeside, to the east of the Cairngorm massif. Grouse feed almost exclusively on heather, and the number of birds is controlled primarily by food supply. Young heather is far more nutritious than old. Good management practice therefore requires the burning of strips of heather so that each territory contains both young plants for feeding and older more woody plants for nesting. The pattern of burned heather strips to the right of the road is typical of a well-managed moor, with a large unburned area to the left. On a larger scale, geology is a major factor in the productivity of grouse moors because of its influence on the growth of the heather itself, and it is generally true that moors further south in England are more productive than those in Scotland. Despite the firm belief (now happily passing away) of gamekeepers, predation by birds of prey is relatively unimportant in controlling grouse numbers, since there are always surplus birds which fail to find a territory. Despite this, birds of prey, including Golden and White Tailed Eagles, are still found poisoned and shot in Scotland.

UPLAND LAND USE CHANGE

Some of the most useful historical data come from the Upland Landscapes Study carried out by the Countryside Commission between 1975 and 1980,[27] although unfortunately the study is restricted to England and Wales. Twelve parishes were selected (the same as those subsequently used by the Institute of Terrestrial

12 *opposite* Oxendale, Langdale Valley in the Lake District, with the broad channel of the river and the boulder trap built to control downstream deposition (and hence flooding). Improved fields on low ground give way to larger intakes now mostly abandoned. The key feature here is the extent of bracken encroaching on what was once pasture. Bracken invasion is relatively slow, taking place by vegetative spread, but is difficult to stop. Chemical treatment or repeated cutting several times a year for several years is necessary to control it. It was once used for bedding. Stock will not eat it (sensibly, since it is poisonous), but trampling by closely stocked cattle damages the fronds as they emerge.

Ecology in their ecological work) in counties ranging from Northumberland to Devon. The areas are extremely diverse, and the representativeness of so small a sample might be questioned: the high moorland of Alwinton in Northumberland is very different from the milder and more lush land of Lynton on Exmoor for example. Land use was mapped at three arbitrary dates, 1872 (a nominal date based on Ordnance Survey maps ranging from 1848 to 1887), in 1967 (using the Second Land Utilisation Survey) and 1978 (from field survey). This is not entirely satisfactory because it soon becomes clear that land was moving more or less continuously into and out of the different land uses throughout the period, especially in the huge gap between 1872 and 1967. There is no record of this turnover or 'flux' in land use, a fact of great importance given the slow ecological recovery of abandoned land.

Nonetheless, it is possible to make useful calculations of aggregate net changes in land use between 1872 and 1978. There was a 12 per cent decline in the area of semi-natural vegetation, a 4 per cent rise in the area of crops and grass, and a dramatic 178 per cent rise in the area of woodland and forestry.[28] There was also a major rise of 127 per cent in the area of land lost to other uses, including water supply, developed land and quarries. Most of the loss of semi-natural vegetation took place between 1967 and 1978, all but one of the parishes showing a decline in this period. Of the 3,629ha of semi-natural vegetation lost between 1967 and 1978, 9 per cent was lost to development, 35 per cent to crops and grass (i.e. intensified agriculture) and 56 per cent to forestry and woodland. Much of the new woodland was coniferous, the proportion of which rose to 84 per cent of total woodland area by 1978. In more detail, there was a marked decline in the area of better bent/fescue grasslands and to a lesser extent in heaths, coarse grasslands (*Nardus* and *Molinia*) and sedges. Bracken and more dramatically gorse expanded. The overall picture is therefore of a loss of ecologically diverse communities, and also a decline in open landscapes. This seems to be the result of a shift away from traditional forms of stock husbandry towards more intensive farming, and away from farming towards other land uses such as forestry. Both the environmental and socio-economic implications of these changes are important (photo 12).

These data from the Upland Landscapes Study are confirmed by studies by researchers at the University of Birmingham of land use change between approximately 1950 and 1980 in five National Parks and in mid-Wales.[29] These cover a far larger area than the 12 parishes study and use air photographs to give data on both temporary and permanent changes in land use. In aggregate there was a net decline of 16 per cent in the area of rough pasture between 1950 and 1980. This was greatest in mid-Wales (28 per cent), followed by the North York Moors National Park (25 per cent). A total of 84,000ha were converted between 1950 and 1980 on a more or less permanent basis, discounting short-lived conversions. This represents 12 per cent of the total study area, and compares with a small area of 1,500ha (2 per cent of the study area) of rough pasture reverting from agriculture or forestry. Of the rough pasture lost, 28 per cent went to farmland and 70 per cent to forestry. Losses to forestry were particularly important in the Snowdonia and Northumberland National Parks.

Some of the pressures on open upland landscapes are subject to planning con-

trol (see chapter 4). One of these is mineral extraction. In some areas, most notably perhaps in the Peak District National Park, mineral extraction is an important and long-standing land use with major implications for maintaining local employment. It also has severe landscape impact. Open-cast mining for fluorspar and limestone in the Peak District is highly controversial, and highlights the weaknesses of the planning system. The Peak District was in fact the first National Park established, and has the only planning board with an independent budget. Other National Parks may have even less freedom within which to plan for mineral extraction as for example in the case of a proposal to mine copper by open cast methods in the Snowdonia National Park in the 1970s.[30]

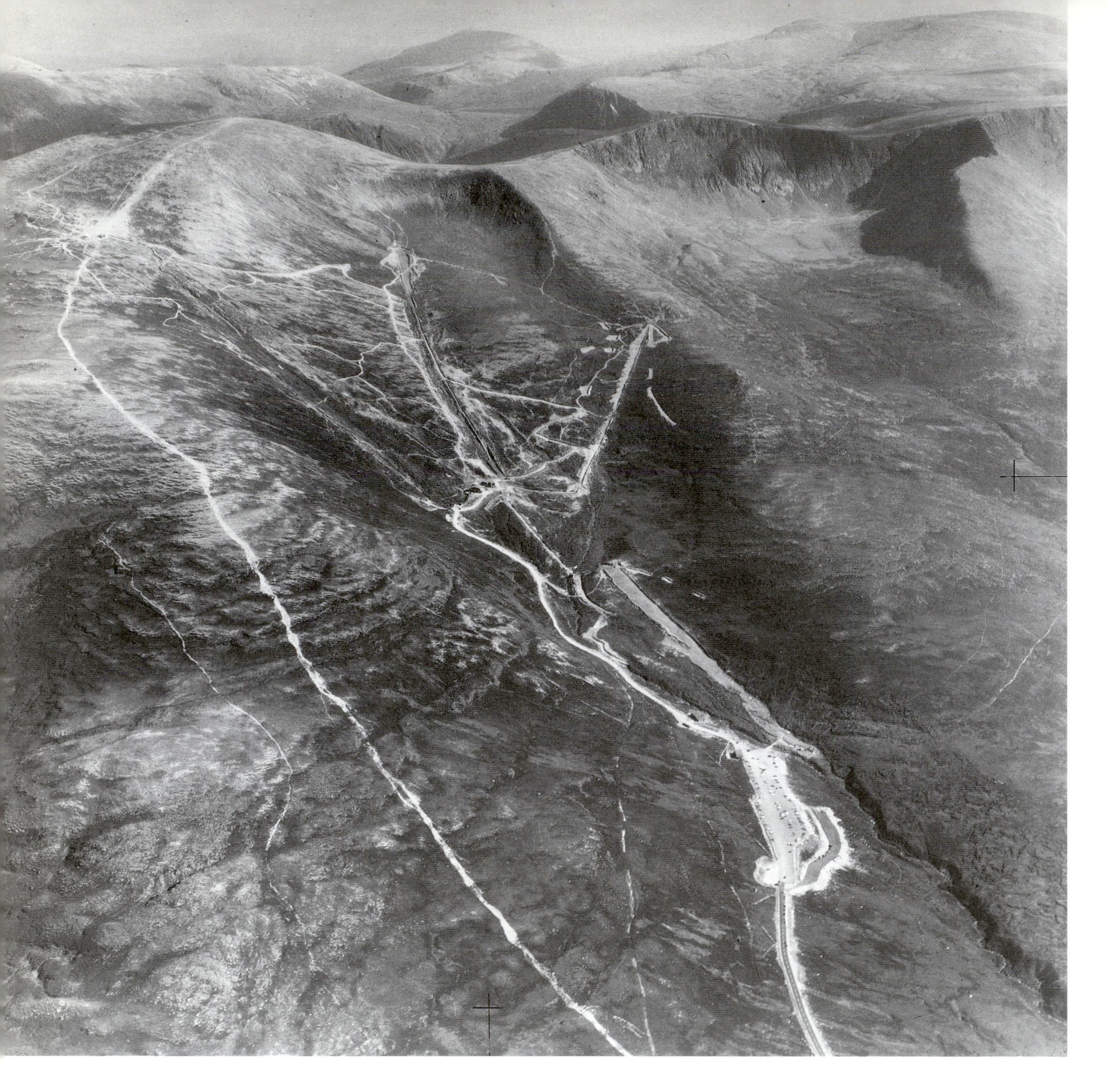

There is also pressure on the uplands from demand for recreation facilities (see chapter 4). There are local erosional problems due to the pressure of walkers on popular mountains such as Snowdon and parts of the Pennine Way. These can be difficult to solve because the environment is harsh and, as a result, the re-establishment of vegetation is slow. Problems on a larger scale are presented by demands for new or expanded downhill skiing areas (photo 13). A long-running Public Inquiry found against a new ski area in Lurcher's Gulley adjacent to existing facilities on Cairngorm, but there are new proposals for a road to this area, and development has occurred in the Lecht Ski area near Tomintoul in the Cairngorms. Ski developments have also been proposed on a number of other mountains including Aonach Mor near Ben Nevis and Ben Wyvis. Early development in the Cairngorms led to major problems of erosion on pistes and around chairlifts. This can now be minimised by careful reseeding with selected grass mixes and fertilisers, but obviously natural vegetation is destroyed. The scenic impacts of ski developments are also considerable, and they will inevitably remain controversial.

Another highly controversial land use is military training. There are live firing ranges in four National Parks, of which three (Dartmoor, Peak District and Northumberland) are in the uplands. In Northumberland, for example, the Ministry of Defence owns, leases or has rights over, about 87 square miles or 22 per cent of the park, and there is live firing on 260–290 days a year.[31] Such military uses are opposed strongly by local non-governmental pressure groups such as the Dartmoor Preservation Association, but demands for such areas are, if anything, continuing to rise (photo 14).

Military uses generally fall outside the planning system, but other development threats do not. Opinions differ on the effectiveness of planning as a way of conserving landscape. Some view National Parks, Areas of Outstanding Natural Beauty, National Scenic Areas and SSSIs as a mothballing of the working landscape. The NCC and Countryside Commissions are accused of unwarranted interference with much needed local enterprise and jobs. These views were forcibly expressed at the Public Inquiry into the creation of a North Pennines AONB in 1986, although the Countryside Commission stressed the additional availability of grants for appropriate developments if designation went forward. Other critics argue that in fact these designations make little difference to the way decisions are made about planning applications: research in Devon, for example, showed no statistical difference between the number of applications rejected within an AONB and outside.[32]

Important though the question of the limitations of the planning system is, it can affect only one element in the equation of landscape change in the uplands. Neither agriculture nor forestry are subject to planning control, yet these are the chief causes of land use and landscape change in the uplands in Scotland as well as England and Wales. Patterns of both upland farming and afforestation are largely the product of broad government economic policies. The impact of these policies on the landscape is largely inadvertent, but none the less important for that. There is a serious lack of integration between the economic, social and environmental policies in the uplands.

UPLAND AGRICULTURE

Hill farming was assisted in the post-war years under the Hill Farming Act 1946. Provisions included grants for capital improvements such as farm roads, fencing, land improvement, farm buildings and homes. Headage payments were also made for sheep and beef cattle. These were intended to offset the handicap of environment and location in harsh and remote hill regions. In this respect they were rather successful, and stocking levels on hill land and the productivity of upland farm businesses rose in many areas. Policies were continued following Britain's entry into the EEC in 1973, and the 1975 White Paper *Food From Our Own*

13 *opposite* Ski lifts and runs of Cairngorm. The specially built road above Glenmore and the main car park can be seen in the bottom of the picture. This has subsequently been greatly extended, and a major lift station and leisure area built at its upper end. From this the access track winds up to the restaurant and ski station at the mid point, and on up to the Ptarmigan Restaurant at 1090m. The summit of Cairn Gorm (1245m) lies behind and to the right of the restaurant. From Cairn Gorm a ridge dips above Coire and t-Sneachda to Cairn Lochan (1215m), behind which stretches the extensive Cairngorm Plateau. A network of ski tows and runs covers the corrie below Cairn Gorm, and the top of the Coire na Ciste runs are just visible below the top restaurant on the extreme left. The most obvious features are the scars left by tracks and tows. To an extent this has been overcome subsequently by planting and fertilising grass, but while this has reduced erosion, eroded areas are still visible on the ground by their bright green grass. The landscape impact of the development is obviously enormous, and significant areas of montane vegetation have been damaged by engineering works or trampling. The ski lifts have also increased summer walking pressure on the Cairngorm Plateau. Despite a decision against further development of the ski area at a Public Inquiry, pressure from economic interests, skiers and those concerned about employment is high. Snow fencing has been put in and an access road allowed further east below Creag an Leth-choin (in Lurcher's Gulley), and further lifts and bulldozed runs seem to many observers to be inevitable.

Resources envisaged continued rising production of meat from upland areas. At this stage agricultural policy in the uplands as elsewhere placed emphasis on raising production above all else.

This attitude was out of step with the EC's Less Favoured Area Directive of 1975, which was based on a social as well as a purely economic role for agriculture, and also acknowledged the importance of conservation. The Ministry of Agriculture (MAFF) and Scottish Department of Agriculture have been slow to adopt this modified emphasis in any meaningful way. Successive pieces of legislation have gradually strengthened the duty laid on MAFF to place conservation on an equal footing with production. Section 11 of the Countryside Act 1968 required all ministers to 'have regard to the desirability of conserving the natural beauty and amenity of the countryside'; Section 412 of the Wildlife and Countryside Act 1981 made the offer of conservation advice to farmers obligatory, and although a clause requiring MAFF to further nature conservation was excluded from the Wildlife and Countryside Amendment Bill during debate a similar provision was subsequently included in the Agriculture Bill 1986.

Whatever the future, British agricultural policy from the late 1940s promoted production in the uplands through grants and subsidies. Payments of all sorts favoured the larger farmer and the less severely handicapped land, and were substantial. In 1980 Hill Livestock Compensatory Allowances (HLCAs) in the United Kingdom amounted to £75.7m, EC and British capital grants to £51.8m, and Guidance Premiums to £3.5m.[33] Of this expenditure, 41 per cent was in England and Wales. On average this amounts to £2,500 per farmer, but distribution is inequitable. Over half the LFA farms in England and Wales in 1981–2 (11,200 farms, 54 per cent) had less than 50 livestock units eligible for HLCAs, and they received only £6.6m, 15 per cent of the total HLCA payments made. By contrast, 759 large farms of over 300 livestock units received £10m, 22 per cent of the total. This latter figure represents over £13,000 per farm, almost three times the average, and 22 times the payments to the poorest 54 per cent of farmers. If extended to Scotland, this inequality would without doubt be greatly increased. Despite substantial payments, or rather because of the single-minded focus on increasing production which determined *how* they were paid, the uplands have undoubtedly seen real economic hardship in the agricultural sector, and associated social and economic decline. Indeed, in a number of places farming has ceased to be the economic mainstay. In the Lake District, for example, revenue from tourism is now four to five times that from agriculture (see photo 58).

Changes in national agricultural policy and EC and Common Agricultural Policy regulations obviously have significant implications for upland land use. Impacts are hard to predict in detail, partly because of the uncertainty felt by the farming community. However, already the imposition of milk quotas in 1984 has seen a number of farmers (6.8 per cent in 1987, 2,500 farmers) selling their quotas. Many have gone over to sheep and breeding ewe numbers grew by 20 per cent between 1983 and 1987. Hill farmers who cannot compete with new sheep farmers on lower and more productive land are likely to face considerable economic difficulties in the coming years, and substantive land use changes may result. They may include the abandonment of marginal grazing land, or conversion of grazed land to conifer forestry.

14 *opposite* The dramatic scenery of High Cup Nick in the North Pennines, looking northeast across the Pennines towards Upper Teesdale and Weardale. The villages of Hilton and Murton in the foreground lie in the fertile Vale of Eden just east of the town of Appleby. To the right of the picture lies Warcop Fell, part of an extensive army range. This area lies within the North Pennines Area of Outstanding Natural Beauty (AONB), created in 1986 after a Public Inquiry. AONBs are designated by the Countryside Commission under the Countryside Act 1968. They provide enhanced access to Countryside Commission grants to maintain amenity and access, and confer some protection in that their special status is more widely recognised. As in the National Parks, military use can be extensive, and sometimes damaging. In the background the black expanse of the Cow Green reservoir can be seen. This is shown better in photograph 9.

THE CASE OF EXMOOR

Until now, of course, it has been agricultural intensification that has had the greatest impact on upland landscapes through the enclosure of rough grazing, and pasture improvement through land drainage, fertilisation and reseeding. This has brought it into conflict with conservation, particularly where conditions for agriculture are least harsh at the lower margin of the uplands. This conflict is best illustrated in the case of Exmoor. Large areas of open moor were enclosed and reclaimed in the nineteenth century, and most could support improved pasture. The parish of Lynton in the Brendon Hills, one of the 12 in the Upland Landscapes Study, is fairly typical: no land is above about 425m, 60 per cent is improved grassland and 16 per cent is tillage.

The story of the conversion and conservation of open moorland on Exmoor is complex. Briefly, the tale begins in 1962 with a proposal to enclose 89ha on Countisbury in Devon in an area of renowned scenic beauty just east of Lynton (photo 15). The two County National Park Committees failed to stop this, Devon County Council refused to acquire the land by compulsory purchase, and the Minister of Housing and Local Government did not apply an Article 4 Direction to make fencing subject to planning control. In 1966 the Exmoor Society published a study of moorland loss, *Can Exmoor Survive?* by Geoffrey Sinclair, which reported substantial shrinkage in the area of open moor from 24,000ha in 1957 to 20,000ha in 1966, mostly through agricultural reclamation. In 1969 the National Park Committee defined a 'critical amenity area' of less than 18,000ha within which they would not wish to see open land enclosed. The 1968 Countryside Act created the possibility of management agreements with landowners to main-

15 Lynmouth on the North Devon coast. The mouth of the Lyn River was completely re-engineered following the disastrous floods in 1956, when a freak rainstorm in the catchment of the river on Exmoor caused major flooding and loss of life. Above Lynmouth the road runs up Countisbury Hill and east towards Porlock Bay. The enclosure of 89ha of open moorland east of Countisbury in 1962, and the failure of either of the two National Park Committees to stop it or the Devon County Council to protect the land by compulsory purchase, was one of the triggers to the production of the report *Can Exmoor Survive?* by the Exmoor Society. Following this the issue of moorland loss became highly controversial, leading eventually to the Portchester Report and new provisions to protect open land in National Parks in the Wildlife and Countryside Act 1981.

tain access to open land, and notices requiring six months notice of intended conversions of moorland, but in practice a voluntary notification system was introduced on Exmoor instead. Despite this, piecemeal conversions continued, until proposals in 1976 and 1977 for the reclamation of Yenworthy and North Commons on the Glenthorne Estate, and nearby Stowey Allotment brought matters to a head. Eventually the Countryside Commission reported the Exmoor National Park Committee to the Secretary of State for the Environment for mishandling the negotiations about the future of this land.

In 1977 Lord Portchester presented his report on moorland change, *A Study of Exmoor*. This confirmed the extent of moorland loss (open moorland reduced to 19,000ha by 1976), and proposed a system of compulsory Moorland Conservation Orders allied with compensation for farmers and landowners prevented from developing land. These proposals were incorporated into the 1979 Countryside Bill which was lost when the government changed. Revised proposals became part of the Wildlife and Countryside Bill introduced into Parliament in 1980, which also contained detailed new provisions for the protection of Sites of Special Scientific Interest (SSSIs). Debate was lengthy and intense, and the resulting Act is extremely complex. It makes provision for the protection of SSSIs (Sections 28 and 29), for the creation of maps of critical moorland areas in National Parks (Section 43), and allows the Secretary of State to make Moorland Conservation Orders to prevent moorland conversion (Section 42). The provisions for moorland maps were subsequently revised in the Wildlife and Countryside (Amendment) Act 1985.

The Nature Conservancy Council (NCC) and the Countryside Commission have suggested that a full decade will be needed to determine the success of these provisions. The NCC had completed the notification of 83 per cent of SSSIs by the end of March 1987, at which point the Act's provisions became effective on these sites. Meanwhile, threats to upland SSSIs and to upland landscapes continue, and it is not clear that the cumbersome and bureaucratic procedures of the 1981 Act will successfully end SSSIs loss and guarantee the conservation of upland landscapes. There are certainly questions over the provisions for landscape conservation, especially in areas outside the National Parks and Areas of Outstanding Natural Beauty. There has also been considerable controversy over National Scenic Areas in Scotland, and the environmental impact of developments under Integrated Development Programmes in the Outer Isles, Orkney and Shetland. The single most thorny issue in the uplands, however, is that of the impact of forestry.

UPLAND FORESTRY

Britain produces about 3.8 million cubic metres of wood per year, 8.9 per cent of requirements, on about 9 per cent of the total land surface. Despite the relatively large areas of broadleaved woodland, some of it in upland areas and much of it held by private landowners (0.5 million hectares, including coppice), most wood production is softwood from conifer plantations. Of the two million hectares of productive woodland in Britain in 1985, 71 per cent was coniferous. Over half the privately owned woodland is coniferous, and most of it is found in Scotland.

Conservationists and foresters endlessly debate the pros and cons of afforestation with conifers. There are certainly both gains and losses. Drainage of boggy land has particularly serious impacts (photo 16). Land preparation, particularly deep ditching, brings a rapid pulse of erosion and nutrient loss while trees are being established. Following drainage, peatland plant communities are replaced by coarse grasslands, and the breeding habitat of various rare birds, particularly waders such as dunlin and greenshank, is lost. Furthermore, planting of even a small area of peatland can affect the hydrology of the whole, and have significant impacts on large areas outside the planting limit. The loss of blanket bog is most serious and most controversial in the Flow Country of Caithness and Sutherland

in northeast Scotland, although serious losses to afforestation have also occurred in the Southern Uplands, Cheviots and Wales.[34]

The planting of open hill land transforms the environment greatly. On drier moorland vegetation can become luxuriant in the first few years of a plantation's life while trees remain small. Birds of open moorland such as golden plover and red grouse disappear as their feeding areas and nesting sites are lost, but others such as the black grouse and stonechat are favoured, as are mice and voles. Trouble comes after 12–15 years when the plantation reaches the thicket stage and the young trees cover the ground and exclude light. The ground flora now declines sharply, both flowering plants and mosses virtually disappearing except on the edges of rides and in cleared areas for the next 50 years until the trees are felled. The bird population of the maturing plantation is quite large in numerical terms, but it is almost made up of common species such as the chaffinch, blackbird and woodpigeon. Almost all the species of the original, open moor are lost. Birds of prey are particularly badly affected. Peregrine falcons, buzzards, merlins and golden eagles all can survive the early stages of a plantation's life, but disappear once the closed canopy stops them hunting. Because these birds need large territories, even quite small areas of planting can deny them part of their hunting area, and cause them to disappear from large tracts of mountain and hill. This effect has been demonstrated for example in the case of the golden eagle in southern Scotland,[35] and it is also a factor in the decline of the merlin.[36] Deer and sheep can be similarly affected if plantations prevent them reaching valley bottom grazing areas in winter. Large expanses of plantation, of course, have greater effect, and studies are increasingly suggesting serious impacts on the acidity of rivers draining forest areas and their fish and aquatic fauna. Soil erosion, effects on hillslope hydrology, acidification of surface waters and its impact on fish, and the effects of fertilisers and pesticides applied from the air are also of increasing concern.[37]

To an extent of course, the adverse conservation implications of afforestation can be offset by good forest design and management (photo 17). Damp areas can be left unplanted, a border can be left along rivers and streams, and the age-structure of the forest can be diversified so that there is a patchwork of habitats throughout its life. Measures of this kind are well understood, but unfortunately they run counter to the pressures for profitability both within the Forestry Commission and in the private sector. Brashing maturing trees encourages the ground flora, but it costs money because it is hard to mechanise. If planting is grant-aided by area, it may be worth planting boggy areas even if trees are unlikely to prosper, and certainly planting is moving higher on to more exposed sites where windthrow is a problem. To counter this, plantations are not thinned, and are cut on a short rotation of 40 years or less. Such timber is of low value, and it creates a forest of even-aged stands, little wildlife value, and maximum landscape intrusion. The integration of conservation and forestry is technically possible, within strict limits and assuming it can be guaranteed that key areas are left unplanted, but forestry's policy framework effectively prevents it on both Forestry Commission and private plantations.

The Forestry Commission was established in 1919 with the primary objective of creating a strategic supply of timber. Its first land was planted in 1921, and

16 *opposite* Silver Flowe in Kirkudbrightshire. Silver Flowe is a SSSI (National) listed in the *Nature Conservation Review*. It is most important for its bog communities, and the pattern of pool and hummocks can be seen to the right of the stream. These bogs form part of a continuity of vegetation communities up onto the open hill to the right. The peats provide an important record of vegetation change since the end of the glacial period through the analysis of pollen. Also visible is the damage afforestation does to such peatlands. To the left of the stream land has been ploughed for planting, and a new road built. Large areas of peatland, as well as open hill, have been planted in Galloway. Once planted, such areas retain almost none of their wildlife interest. Planting on even part of a bog system can damage the hydrological processes on which the whole depends, and cause damage well beyond the limits of planting. Such problems have become serious on a large scale in the Flow Country of Caithness and Sutherland.

by 1929 it had planted 56,600ha. By 1949 it held 0.63 million hectares, of which 225,000ha were planted. Most of that land was in the uplands, and most was planted with conifers. Forestry Acts were passed in 1945 and 1947 which established the FC on a firmer footing both as a Forest Authority (responsible for the whole UK forestry industry) and a Forest Service (itself holding land and growing timber). During its life, its role has been redefined several times. By the 1960s the strategic justification for forestry had become more or less irrelevant. In 1972 the economic soundness of large-scale planting was questioned, and other benefits such as recreation and (as in the 1920s and 1930s) forestry's role in generating employment became relatively more important. The pendulum was pushed back again in the late 1970s, and by 1980 optimistic economic forecasts were once again being made by the industry. The FC itself however was required by the

17 Afforestation near Rochester in the Cheviots, a few miles south of the border. In the foreground runs the A68 from Newcastle to Edinburgh. The sharp edges of the plantation and the cuttings made for access roads within the forest can be clearly seen. The planting forms part of the Redesdale Forest, an extension of the vast Kielder estate acquired by the Forestry Commission soon after its formation in 1919. The Cheviot Hills are bleak, and their soils shallow and sour. Few timber species will grow, and the Kielder Forest is almost 100 per cent Sitka Spruce (*Picea sitchensis*), a grey-green monoculture covering thousands of hectares. The danger of windthrow in strong winds means that the trees have to be harvested young (30–40 years), which makes for poor economic returns. Also, they cannot be thinned. The resulting plantations are dark and impenetrable. Much of the forest is still in its first cycle of growth, and the blanket of trees contains large areas of uniform age. As these are cut, attempts are being made to plant smaller units which will in time create a more diverse age structure. Corridors and patches of broadleaved trees are also being planted. The resulting forest will be far more attractive, with enhanced recreational and wildlife value.

18 *opposite* Forestry surrounding the Llynn Brianne reservoir in the Tywi Valley north of Llandovery. The area was forested with exotic conifer species before the reservoir was built. The layout of forestry roads, and the contrast between the plantation and the bare mat grass (*Nardus stricta*) dominated grazing land behind are clear from the photograph. Forestry and water supply are two of the land uses which compete with extensive grazing for the relatively harsh and unproductive landscapes of areas such as central Wales.

government in 1981 to reduce its estate (21 per cent of which was still not forested in March 1985) to release excess capital. In 1985 the FC's budget stood at about £150m, of which only a third was covered by timber sales. Sales slowed after 1984, but rationalisation of its estate continued (photo 18).

Private forestry has fared rather better in recent years. Government assistance to the private sector began with the Dedication Scheme (introduced in 1947) and the Approved Estates Scheme (1953). The introduction of Capital Transfer Tax in the late 1970s cut back the rate of private planting to 10,000ha per year, but recently it has expanded again. In the year ending March 1985, 19,000ha of planting was grant-aided in England and Wales, 91 per cent of it with conifers, and

19 *opposite* Peaks of the Brecon Beacons.

80 per cent in Scotland. In the year 1983–4 total FC grants to woodland owners, most of which were for planting, amounted to £5.8m. The Forestry Grant Scheme (FGS) in Scotland in 1984–5 was probably worth £3.32m, of which conifers made up 97 per cent.

However, by the mid-1980s, it was becoming clear that the greatest element of government support for private forestry was not direct grants, but indirect tax benefits.[38] Until the system was changed in the 1988 Budget, anyone paying a high rate of income tax could treat the cost of establishing a forestry plantation as a tax-deductable allowance. Those who earned large sums over short periods, such as professional sports competitors and entertainers, could capitalise their income and offset the expense against their total tax liability. A number of consultancy and management companies such as Fountain Forestry, the Economic Forestry Group and Tilhill worked in this field.

The benefits of private forestry investment under this taxation regime could be considerable. Planting on typical hill land might cost £1,000 per ha. Under the FGS (at the rates revised in October 1985) grant could be paid of £240 per ha against establishment costs (fencing, roads etc.) on areas of over 10ha. This brought the net cost of establishment down to £760. Under Schedule D of the Income Tax regulations, it was possible to offset 60 per cent of this against tax, bringing the net cost down to £304 per ha. The total public subsidy, including tax, was £696 per ha. However, the tax benefits to the private forester did not end there. Once the plantation was established, annual costs fell and the advantages of ownership declined. It was then possible to sell the land (perhaps to a close relative, or a company established for the purpose) at which point taxation could switch back to Schedule B. Under this, tax was paid on one third of the unimproved value of the land, but not on the timber itself. The timber could later be sold tax free. Tight financial management meant that very little money went back to the public purse at any stage, and the investor gained all down the line.

The support for private forestry from taxation came under considerable pressure from environmental groups in the mid-1980s, generating controversy in places (notably the Flow Country) where forestry jobs were a significant element in employment. This tax regime changed in 1988 to remove the incentive to high-rate taxpayers. However, the Forestry Commission announced changes in grants procedures which more or less completely compensated for the tax cuts. While this new regime of grant aid had many environmental advantages, making the planting of small farm woodlands more attractive, for example, it seems unlikely to bring any diminution in the pressures on upland planting.[39] The Secretary of State for the Environment announced a halt to planting in the National Parks in England and Wales (photo 19), but planting in Scotland (without National Parks, of course) will persist. The Secretary of State for Scotland announced that the target of 33,000ha of new ground planted every year in the UK would be maintained. In the year to March 1988, the Forestry Commission planted 4,995ha and grant-aided 23,821ha of private planting. Pressure on the Flow Country has been maintained, and further planting has taken place.

Unsurprisingly, the area under private forestry has expanded considerably in recent years. Applications for Forestry Commission Grants rose from 2,000 in

1984 to 5,362 in 1987. The area of private planting in Scotland in the year 1984/5 was 14,100ha. One company, Fountain Forestry, own or manage somewhere between 20,000 and 30,000ha of land, of which they are planting 12,000ha, mostly with sitka spruce and lodgepole pine. Much of the land lies in the far north in Sutherland and Caithness. In this environment there must be doubts as to whether the trees will grow well enough to be economic on the poor peaty soils without suffering windthrow damage. There are fears, too, about the risk of attack by pine beauty moth and the great spruce bark beetle in extensive plantations of a single species. Epidemics elsewhere have required costly spraying from the air with the general insecticide fenitrothion, and other compounds. The ecological impact of this treatment is not understood in any detail.

There are no legal controls on tree planting, and despite its implications for both landscape and nature conservation it is not subject to planning control. Instead, would-be foresters have been expected to ask the FC for approval. There is more force to this than might at first appear because unless approval is given the grant will not be paid. However, forestry has become so attractive to investors that demand for land with planting approval has forced its price well above (indeed up to twice) that for ordinary hill land. That difference more than cancels out the value of any lost planting grant. Furthermore, there is no limit on the area which may be planted, and the whole area remains eligible for tax relief. Such planting without approval (which is not illegal in any way) is obviously very attractive. In April 1985 planting began on an 800ha property in the Lammermuir Hills south of Edinburgh without approval. A month later a further 200ha was planted by a private forestry company at Shielsknowe near Carter Bar in the Borders region.

These controversial actions generated a sharp debate in the Scottish Press. Conservation groups such as the Royal Society for the Protection of Birds, Friends of the Earth (Scotland) and the Scottish Wild Lands Group are unhappy with much of the *agreed* planting, and are fearful of the implications of uncontrolled development, and the Department of Agriculture is worried about implications for hill farming in Scotland. The Borders Regional Council called in August 1985 for planning controls, something which an increasing number of observers have been advocating.

The Director General of the Forestry Commission claimed reassuringly in 1986 that there was 'no free-for-all' in Scottish forestry, and certainly the replacement of tax benefits with grant aid in 1988 places the initiative for controlling forestry firmly back in the Commission's hands, because it is now uneconomic to plant without it. However, it is far from clear to what extent the Commission will use this new power to direct planting away from key conservation areas.

One bright spot is the Commission's announcement in 1988 that an 'Environmental Assessment', under the European Community's Environmental Impact Assessment Directive, would be required for some planting applications, particularly those affecting SSSIs. In 1988 the need to integrate forestry sensibly into upland use policy remains very great.

One of the key questions is certainly that of the future of Sites of Special Scientific Interest (SSSIs) in areas threatened with planting. In theory, the Wildlife and Countryside Act 1981 means that the Nature Conservancy Council is informed about all possible damage to SSSIs, and it has stated that it will offer management agreements to compensate landowners who want to plant on SSSIs. The trouble is that the calculation of the benefits of forestry enterprises, with the need to discount costs and benefits over the 50 or so years of a plantation's life, is exceedingly complicated, especially in cases where the *real* benefits relate to overall tax liability and not the value of the trees or the land. There is little experience as yet with such agreements, but the indications are that foresters do not find them attractive. In this circumstance, landowners can simply turn them down. The only way the NCC can stop the planting is to ask the FC to withhold planting approval. But, of course, even though grant aid is now very valuable, the forester does not legally *need* that approval. Despite the 1981 Act, the NCC cannot guarantee the protection of SSSIs. It can only *ask* landowners and the FC to co-operate. If the proffered management agreement is turned down, the NCC's only recourse is to ask the Secretary of State for a Compulsory Purchase Order. This is a cumbersome and heavy-handed procedure, and is politically sensitive. Certainly controversy remains over new planting. Only 0.3 per cent of applications for planting grant were referred to Regional Advisory Committees in 1987.

The bizarre economics of afforestation create a number of problems for the conservation of upland SSSIs. The most publicised case of conflict between afforestation and nature conservation, over Creag Meagaidh in the Scottish Highlands, gives little encouragement. Creag Meagaidh is a mountain beloved of climbers rising high above Glen Spean. It forms an SSSI embracing continuous vegetation communities from the floor of the glen (partly flooded by a reservoir) to the summit around 1,200m up. In 1984, Fountain Forestry bought 4,000ha of the estate for £300,000 (£75 per ha) and applied for approval to plant an extensive area. The NCC objected, and after extensive debate with the FC, the question was referred to the Secretary of State for Scotland. He proposed a compromise, whereby Fountain could plant part of the site (530ha). This was unsatisfactory to both parties; too little for Fountain, but still too much for the NCC, and the dispute rolled on. Eventually, after tussles between the Secretaries of State for the Environment (under whom the NCC came) and for Scotland, Fountain were persuaded to sell the whole site to the NCC for £430,000 as a National Nature Reserve. All may be well that ends well: Fountain made a substantial profit, the NCC protected the site. Nevertheless, Creag Meadaidh is hardly a hopeful model of land use planning.

In June 1986 the NCC produced a forthright policy paper, *Afforestation and Nature Conservation in Great Britain* This stated the case for conservation in mountain and upland communities, and the threat of uncontrolled afforestation, in unequivocal terms. It called for a halt to all planting on SSSIs, and for clear statements from the FC on its plans to implement the new duty laid on it in the Wildlife and Countryside (Amendment) Act 1985 to 'achieve a reasonable balance' between conservation and forestry. Its strictures represent a significant tightening of official attitudes on upland afforestation, and seem to have been taken on board to some extent at least by the government. Their paper *Conservation and Development: the British Approach* recognised that 'very real issues, particularly in areas of high conservation interest attractive to forestry, will require frank debate in the next few years'. This debate has to an extent begun. Its outcome will be

of great importance in determining the nature of upland landscapes in the future (photo 20).

The ways in which policies concerning forestry, as well as other aspects of upland land use, will develop remain as yet unclear. It is obvious that they will be the subject of close and increasingly informed debate within government and outside. Opinion in many quarters now seems to favour a major rethink of policies for the countryside, especially for the uplands, and both government and voluntary organisations are starting to call for this. It is to be hoped that this can be done, and that 'integration' can be made to mean more than orchestrated conflict in which the strongest party (i.e. the one yielding the greatest short-term profit) wins.

20 *opposite* Baosbheinn in Torridon, looking towards the southeast, with the ridges of Liathach and Beinn Eighe in the background. Loch a' Bhealach is to the right of Baosbheinn, Loch na h-Oidhche to the left. The Torridon hills form an extensive wild area, the mountains formed of old red sandstone, most of them topped with resistant quartzite. The result is a series of dramatically steep ridges and deep corries. The area is famous among mountaineers but also has important plant communities in small flushes and mountain meadows. Beinn Eighe itself forms part of a National Nature Reserve, the first declared in Britain, in November 1951. The NNR is managed with the Shieldaig and Flowerdale Forests as a single unit for deer stalking, an important element in the economics of the highlands. Torridon has suffered less than many other areas from construction of access tracks.

FURTHER READING

W.M. Adams, *Nature's Place: Conservation Sites and Countryside Change*, London, 1986.

W.M. Adams and J. Budd (eds), *Monitoring Countryside Change*, Chichester, 1990.

A.R. Clapham (ed.), *Upper Teesdale: The Area and Its Natural History*, London, 1978.

P. Lowe, G. Cox, M. MacEwen, T. O'Riordan and M. Winter, *Countryside Conflicts: The Politics of Farming, Forestry and Conservation*, London, 1986.

A. MacEwen and M. MacEwen, *Greenprints for the Countryside? The Story of Britain's National Parks*, London, 1987.

J. Raven and M. Walters, *Mountain Flowers*, London, 1956.

2 The changing lowlands

The lowland landscape as we see it today has been moulded by human use and exploitation. Originally the lowlands were covered in forests interspersed with tracts of wetland, and this is the condition to which they would return within a few decades if we and our farming were to vanish from the scene. Distinctive and often ancient habitats, such as heathland and chalk grassland, were formed after forest clearance and maintained by use as pasture in various forms within the pattern of traditional agriculture. Fragments of the original forest and wetland survived into modern times alongside these derived habitats.

In this chapter we will be concerned particularly with developments in the last half century or so, a period in which profound changes have been brought about by the modernisation and expansion of agriculture. Driven initially by the need to produce more food at home during the emergency of the Second World War, the impetus behind agriculture was maintained first by the 1947 Agriculture Act, and then by Britain's entry to the EC and its Common Agricultural Policy in 1973. Under these arrangements, agriculture has been supported by guaranteed prices, subsidised by grants, free advice and research, and has been largely unfettered by planning restraints. Every effort has been made to increase mechanisation and decrease employment. As a result, more and more land has been brought under the plough; land has been increasingly heavily treated with fertilisers, herbicides and pesticides; and features, such as hedges and small fields, which were useless, less profitable or simply awkward, have been destroyed.

Post-war changes in the landscape have been so considerable, and so often regarded as 'destruction', that we have come to visualise the pre-war landscape with green-tinted spectacles. This was, however, by no means a natural landscape. Significant changes had been effected by Mesolithic peoples towards the end of the last glaciation, and very extensive changes were brought about by Neolithic farming. Indeed, it is doubtful whether any part of the lowland landscape remained truly natural beyond Roman times, save for the lowland raised mires, which remained essentially natural and used for little more than forms of hunting and gathering into modern times. Whilst most of the land had been shaped by agricultural factors, parts had been retained in a semi-wild state (medieval forests) and other parts had been deliberately created to express a particular view of the relationship between man and nature (eighteenth-century landscape movement). At various times the pre-twentieth-century changes were as profound as those of the last 50 years, notably the enclosure of common pastures and fields in the eighteenth and nineteenth centuries, the drainage of fens, and the deforestation and general woodland clearance of the mid nineteenth century. In some districts cultivation in the early nineteenth century extended as far as it does today. In fact, the half-natural, pre-war landscape which is celebrated in books of late

Victorian and Edwardian photographs represented a recent retreat of agriculture, the product of several decades of agricultural decline and rural poverty famously characterised in Ronald Blythe's *Akenfield*.

Traditional land management was more pervasive than modern land management – every corner was used – but it was not as intensive. However high the area of cultivation became, several elements of stable and conservative use remained. These included the ancient woods, treated as coppice or wood-pasture; heathland commons; extensive pastures on higher, less fertile ground, slopes, and valley plains subject to flooding; haymeadows; and turbaries. Such places were not just stable: they inadvertently preserved a variety of semi-wild habitats within the agricultural matrix.

Traditional landscape was far from uniform. The underlying rock, soil and land forms inevitably influenced the proportion of pasture and tillage. Some areas were cleared of woodland at a very early date, whereas others remained well-wooded into modern times. A major division can be recognised between 'arden' and 'felden' districts which has its roots in pre-history. On top of all this, open field agriculture became strongly established in central England and other districts during early medieval times, and eventually gave way to enclosure and the planned countryside of straight hedges. Elsewhere, the irregular enclosure pattern of pre-medieval farming survived with its scattered settlements as the unplanned countryside.

The following sections quantify some of the changes in the landscape. The emphasis throughout is on the semi-natural elements, those bits which have not been managed by ploughing and its attendant mechanisation, drainage and chemical doses. The statistics refer wherever possible to the lowlands or parts of it, but some of the data were collected at a county or country scale where it is difficult to make a clear distinction between upland and lowland. Although we will be concerned primarily with recent changes, these must be put in their historical context, in order both to demonstrate the nature of the features being changed, and to separate mere fluctuations from profound change.

WOODLAND

Woodland in Britain stood at almost its lowest ebb in 1900, when only 4.5 per cent of the land surface was wood (perhaps 5–6 per cent in the lowlands). In its natural state, almost all the lowlands beyond the coastal fringe and the great mires must have been wooded. Millennia of forest clearance reduced the cover to about 15 per cent by the eleventh century, and significantly further by the fourteenth. During the twentieth century, however, the tide turned, and the national woodland cover has been doubled, so that it now stands at 10 per cent. Most of this increase was in the uplands, but not entirely so. A vast new forest (by British standards) was created on the Breckland heaths, with lesser new forests on heaths and downland elsewhere (photo 21).

Afforestation was by no means an exclusively twentieth-century process. Indeed, though it was probably insignificant in terms of area, some planting was undertaken in the Middle Ages. It gained real impetus in the seventeenth century, so that by 1800 plantations were scattered in small fields, field corners, in belts

(photo 35) and on difficult ground on hillsides, quarries, etc., almost throughout the lowlands. New woodland was also formed by entirely natural processes, as heathland, grassland and meadows were abandoned to natural succession, and grew into scrub and thence to woodland (photos 24 and 31). This process appears to have been particularly common on the southern chalklands in the period between the retreat of the Romans and the arrival of the Normans, and in the period following the Black Death.

The woods of 1900 were not just plantations or scrub on old fields. In fact, the majority were survivors from the Middle Ages, and even earlier, that is, they were ancient woods. Many originated as fragments of the primitive forest cover which was never cleared, but none was left in a pristine condition. They survived because they were useful, managed either as coppice or wood pasture. Coppices (photo 22, top) were woods which were periodically cut down, then allowed to grow up again by means of sprouts from the cut stumps. This produced a thicket of underwood, in which stood a scatter of taller, timber trees – usually oaks – which were grown on as standards. Wood pasture (photo 23), on the other hand, was woodland which was permanently open to grazing animals. Under this regime regeneration and regrowth from stumps was much inhibited, so the tree cover was maintained by perpetuating the existing trees to a grand antiquity. In practice, the tree cover usually became rather open, like parkland, which is a form of wood pasture, and the trees were cropped by pollarding, i.e., cutting off their crowns, and allowing the crowns to grow again out of reach of the grazing animals. Wood pastures were typically associated with forests, parkland, and commons, whereas coppices were associated with private woods not subject to rights of common pasturage.

By 1900 most of the coppices were still regularly cut, but the extensive wood pastures of the Middle Ages had been reduced to small fragments and a few extensive survivals, most notably in the New Forest. Despite steady reduction by clearance over the centuries, the coppices were still scattered over most of the lowlands. Most were small, and few were more than 100ha. Some districts had a wood or two in each parish, and by British standards were well-wooded, but only in parts of the Weald, Chilterns and the southern Welsh borderland did the total woodland area amount to much above 25 per cent of the total land surface (photo 25). Although coppices were well distributed, there remained some conspicuously wood-free districts, such as the Fens, and the belt of clays running to the north and west of the Chiltern scarp.

During the twentieth century, the processes of clearance, afforestation by planting, and natural succession to woodland have continued. Clearance was relatively slow before 1945, but has since accelerated (photo 22, bottom). This can be illustrated by study areas in the east Midlands, where there were 9,429ha of woodland in 1886: in the 60 years to 1946, 555ha were cleared, but in the 27 years from 1946–72 a further 818ha was cleared. The rate of clearance appears to have accelerated during each decade until about 1985, although the present rate seems to be much reduced. Most clearance has been to agriculture, with housing, industry, roads, military installations and even recreation grounds adding to the losses. In some districts the area cleared for ironstone, coal and other forms of mining and quarries has been as large as the clearance for agri-

21 *opposite* Weeting Heath, Norfolk. One of the earliest, and still one of the largest new forests created after the formation of the Forestry Commission in 1919, was on the Breckland heaths. The vast geometric plantations of mainly pine have now grown to maturity, so they are being felled in patches and replaced by second rotation plantations (mid-left). Much of the unforested land was more recently converted to arable, but fragments of the heathland remain, mostly as nature reserves. Weeting Heath National Nature Reserve (centre) is more a grass heath, which has been periodically cultivated in historic times, and latterly has been maintained by rabbits. Today, parts are being colonised by coarse grasses following the reduction in the rabbits, and bracken patches are spreading. Stripes in the heath vegetation, due to underlying soil patterns caused by the last glaciation, are still visible in the reserve, but have been obscured and disrupted by the plantation and the associated ploughing. Nature reserve management makes its mark in the form of a roadside strip which is ploughed as a firebreak.

22 *top* Bradfield Woods, Suffolk. This is a rare survival of coppicing, a form of woodland management which dates back to the Bronze Age in lowland England. This particular wood is managed in much the same way now as it was by its medieval owners, the monks of Bury St Edmunds. The wood is divided by open rides and smaller tracks into many compartments, or 'fells'. Each year, one or more of these patches is cut down, leaving a few larger trees and saplings. The cut trees spring again from the stumps to form the underwood, whereas the retained trees and saplings grow on as standards, or timber trees. The photo shows fells in various stages of regrowth.

bottom Sadly, a substantial portion of the wood was cleared in the early 1970s. The last stages of clearance show the trees and underwood, which have been ripped out, piled and burned, and also the laying of new field drains. Beyond are the new fields, created from woodland in the previous two years: inevitably they have straight edges, though some standards have been retained for 'amenity'. Close inspection would show that the soil is still thick with undecayed trees, roots and twigs. With impressive bravado, the farmer threw open his new fields for public inspection when the remaining wood became a nature reserve.

23 Moccas Park, Hereford. This medieval deer park extends from the flood plain of the river Wye to Dorstone hill to the south. The lower part retains its open park structure, with ancient oaks set in pasture still grazed by fallow deer. Although this can legitimately be regarded as a surviving portion of ancient landscape, there are still elements of change. Much of the pasture has been improved by bracken mowing and fertilisers. The accumulations of fallen branches have been cleared away. Trees planted in the nineteenth century, which would have provided successors for the medieval oaks, were felled in the twentieth century. The parkland's survival depends on the medieval oaks surviving beyond the year 2100, while new successors are planted and allowed to grow to maturity.

The park is set in a landscape of piecemeal enclosure, with hedged fields fitting together in irregular patterns. Small woods within this are linked ecologically through the hedge network. Such patterns have been lost recently from much of the lowlands, but survive well on the upland margins.

24 Thursley Common, Surrey. Surrey contains one of the great concentrations of lowland heath, mostly on commons with light, acid, and infertile soils. The dark, heather-dominated vegetation is relieved by patches of pale wet heath (top), leading down to acid mires in the valleys (out of picture). Originally, this mosaic was maintained by sustained grazing and regular burning, with considerable paring of peat for fuel. Now, however, the commons are rarely used by the right-holders, and they have become a public open space. Although fires still occur irregularly, as in the drought year of 1976, scrub woodland is now spreading. Whilst this mostly consists of native trees, such as oak and birch (centre), pine is also common if there is a nearby seed source (top).

culture. This particular form of woodland loss has been exacerbated by the habit of attempting to hide mines behind trees.

Scrub growth on abandoned heaths and farmland has been appreciable, though the pattern has changed over the years, and some of the scrub has since been cleared. Old commons throughout the lowlands have developed into woodland as common pasturage rights have fallen into disuse. Newly wooded commons are particularly frequent in Surrey (photo 24) and the Chilterns. The post-war arabalisation of lowland agriculture has left many steep slopes both ungrazed and unploughed, so on large stretches of chalk downland, as well as small baulks in large arable fields, scrub of various kinds has developed. Before 1940, the depressed state of farming resulted in extensive thorn scrub growth on unused fields over clay; such scrub has since mostly been cleared, but a few patches remain, some of which are developing into quality ash and oak woods.

Afforestation by planting accelerated after the formation of the Forestry Commission in 1919. Indeed, their earliest, and still one of their most extensive new forests was on the Breckland heaths. Whereas pre-twentieth-century afforestation mainly took the form of small, scattered blocks or belts, afforestation after 1919 was more concentrated into massive new forests, not just in Breckland, but also in the Sandlings, the Dorset heaths, and the West Sussex downs. As a result, huge differences in the afforestation rate emerged between nearby districts: for example, between 1946 and 1972 afforestation occurred at 0.4ha/yr in West Cambridgeshire, but 18.0ha/yr in Rockingham Forest (where restoration of mine spoil was a factor). By 1980, 35 per cent of all the woods in Nottinghamshire had originated after 1920, whereas in West Cambridgeshire only 16 per cent of the 1972 woods had originated after 1886. The types of land which were afforested were those which were least useful agriculturally, for example, in Devon outside the national parks, 81 per cent of the 13,500ha of afforestation between 1905 and 1978 was on rough grassland and heath, and only 19 per cent was on agricultural land and orchards. Lowland raised mires have been severely affected by afforestation. For example, 70 per cent of the moss area present on the north Solway in 1947 was lost to afforestation by 1978.

The effects on the pattern of woodland have apparently been to accentuate the difference between districts. Where woods were few and small, forestry has been insignificant, and there has been net clearance. Where woods were many and large, forestry has been worthwhile, and the amount of new woodland has often exceeded the clearance. With the advent also of new, large forests, the well-wooded districts have tended to become as distinct as they were before the widespread planting of small woods on farmland in the nineteenth century filled in the gaps.

Though the changes in the amount and pattern of woodland have been visually important, the major new ecological element in the situation has been change in the form of management of the old coppice woods. Coppicing was already in decline in 1900, but it was only after 1930 that this practice finally ceased in most coppices outside south-east England. As a result, huge numbers of woods became derelict in the sense that they were just allowed to grow up untouched. The rides became overgrown, and the mixed coppices and their flora were steadily impoverished by the natural processes of competition. The large standard trees had frequently been removed without replacement between 1915 and 1945, and the survivors were engulfed in the upgrowth of the underwood.

Unworked coppices were used to a limited extent as game preserves, but their value to their owners was otherwise slight. Many were cleared to cultivation, but more were cut down and replaced by plantations. Initially, these were broadleaved entirely, or in mixture with conifers, but during the 1950s it became increasingly common to replace overgrown coppice with pure conifers (photo 25). Essentially, the upland afforestation techniques of forestry were applied to lowland coppices, and thereby transformed these woods as features in the landscape.

The fate of ancient woods was closely bound up with the coppice system. In 1945, roughly 500,000ha of ancient, semi-natural woodland remained in Britain as a whole, and in the lowlands the great majority was still, or had recently been, coppiced. Limited areas were treated by other systems, such as wood-pasture or beech high forest. Since 1945, many local studies taken together indicate that approximately 10 per cent has been cleared, 30 per cent has been replanted as high forest, 50 per cent has been neglected, and only 10 per cent remains under traditional management.

25 *opposite* Wyre Forest, Shropshire and Worcestershire. The Wyre Forest is one of the few truly extensive areas of ancient woodland in the English lowlands. Here, on the western side near Cleobury Mortimer, fields have encroached from the north and south, but the forest retains its unity. It is drained by several small streams, which come together as the Dowles Brook, and flow into the river Severn above Bewdley. The forest has been crossed by the branch railway linking Bewdley and Wooferton, which closed in 1962. Later, it was crossed by the water pipeline from the Elan Valley to Birmingham.

Various stages in the development of lowland woods and forestry can be seen. Traditionally, this woodland was treated as sessile oak coppice with birch, diversified by hazel, ash and alder along the base-rich and moist soils of the valleys and streamsides. These coppices were made accessible by a network of winding tracks and rides. A block of such woodland survives (A on accompanying diagram), but like most coppices it has not been cut in recent years. Now it is part of the Wyre Forest National Nature Reserve, and some of the oak is good enough as timber to be a registered seed stand.

Large areas of the coppice have been cleared and replaced by the straight lines of modern plantation forestry. The wide forest roads, which are often straight in upland afforestation, are sinuous here, because they have been forced to respect old ride systems and the lie of the land. Modern forestry is a thing of fashions, here manifested as two types of planting, pure coniferous after clear felling (B), and conifer strips planted into oak coppice (C). The retained oak has since been removed, so the difference was transitory. The forest roads (broken lines) were constructed from imported stone, which was far more alkaline than the acid forest soils: as a result, new species have been introduced, such as deadly nightshade.

The distribution of woodland has not been constant. Plantations on an old field (D) are one of a number of small expansions associated with modern forestry. Indeed, modern plantings have quite obscured the historic boundary between woodland and farmland, which persisted until the early twentieth century. Sturt Common (E) was unwooded in the early nineteenth century, but has reverted to woodland since. Embedded in the forest (F) is a former coppice which was opened to parkland with bracken. Much of this has since been ploughed and grassed. Yet another facet of flux in the wooded area can be seen where the railway crosses fields (G): the abandoned track is now covered in scrub. To the north (H), hawthorn scrub surrounds a pond.

Outside the woods, the fields are a mixture of arable and pasture. Locally, meadows survive with massive anthills and a rich flora (I). Fragments of nineteenth century orchards of damsons and cherries survive (J), a characteristic feature of Wyre.

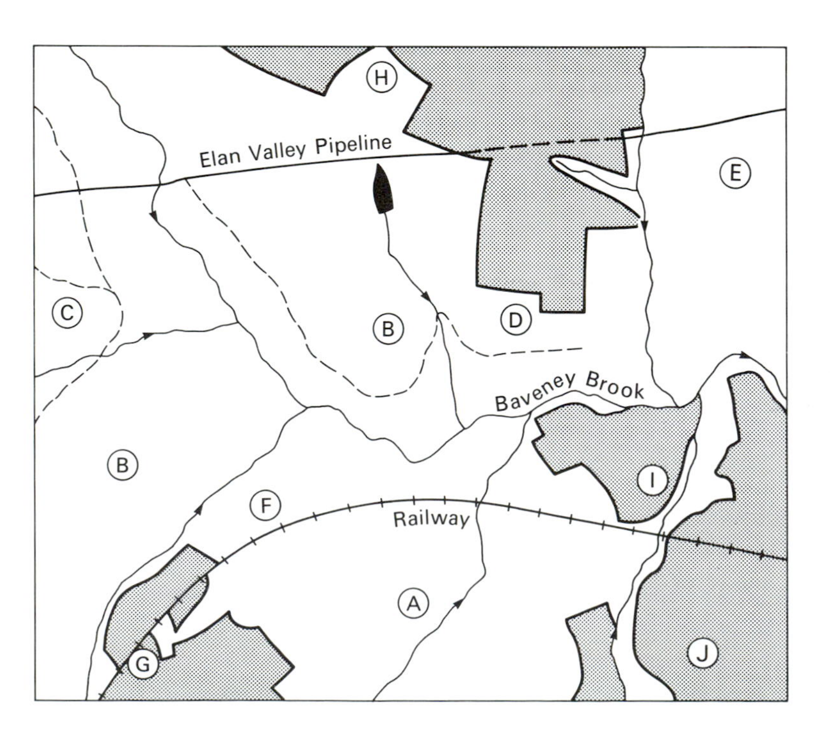

Outline map of the Wyre Forest as shown in photograph 25.

The effects of all these changes on the structure and composition of lowland woods have been profound. Census results for England indicate that the amount of coppice has declined from 30 per cent of all woodland in 1924 to 4 per cent in 1980. Conversely, the extent of high forest has climbed from 46 per cent to 86 per cent over the same interval. Wartime fellings and coppice neglect boosted the proportion of scrub, devastated and felled woodland to 27 per cent in both 1947 and 1965, but this has since dropped below 10 per cent. In round terms, the area of broadleaf woodland has fluctuated from 520,000ha in 1924, up to 585,000ha in 1947, down to 550,000ha in 1965, and back up to 560,000ha in 1980. Correspondingly, the area of conifer woodland has climbed steadily from 140,000ha in 1924 to 388,000ha in 1980. Since these figures include the massive Kielder forest in the Pennine uplands, we can safely conclude that lowland woods remain predominantly broadleaved. Their composition has changed, nevertheless. Changes in Nottinghamshire between 1948 and 1980 included a major loss of oak, and gains for birch, sycamore and beech.

HEDGES AND TREES IN FARMLAND

The boundary between what is and what is not woodland is a good deal less sharp than the Ordnance Survey maps would have us believe. Many of the wood-pastures have trees at such wide spacing that they are not censussed or mapped

26 Wickham Skeith, Suffolk. The village is surrounded by irregular hedged fields, formed presumably by a process of piecemeal enclosure. In the foreground similar fields have been amalgamated, leaving the former hedge-and-ditches as soil marks. Hedgerow trees are concentrated around the village, as is commonly the case in East Anglia and the Midlands.

as woods, even though they are ecologically a form of woodland. More importantly in landscape terms, hedges and hedgerow trees have bounded most of the fields in the lowlands, forming a lacework of woodland habitat between the woods themselves.

Hedges are commonly regarded as recent additions to the landscape, the product of quickset planting when the open fields were enclosed in the eighteenth and nineteenth centuries. This is true of most of the hedges in most of the Midlands clay belt, the core of the planned countryside (photo 26), but even here there are small districts and individual hedges which date from much earlier periods, and which can be recognised by their mixed composition. These include the closes round villages and isolated farms; the former margins of cleared ancient woods, which are often found near the surviving woods and parish boundary; hedges along lanes that have been incorporated into the post-Enclosure field pattern. Outside the planned districts in the majority of the lowlands, the field pattern largely evolved in the Middle Ages and earlier: hedges there are mostly at least 500 years old. The great antiquity of some hedges is attested by numerous references to hedges on property boundaries in Anglo-Saxon charters. Older hedges and enclosure patterns impart a characteristically irregular aspect to the landscape: not only are the boundary lines themselves sinuous, but the fields fit together in irregular shapes.

27 *top and bottom* Madmarston Hill, Swalcliffe, Oxfordshire. Between 1961 (top) and 1968 (bottom) the hedges, ditches, hedgerow trees and the ramparts of a small hill fort were all destroyed in the process of changing the fields from permanent pasture to arable. This was not, of course, the first time the land had been cultivated: ridge-and-furrow shows that the whole area was once part of an open field system, leaving only the land within the ramparts and the field by the stream uncultivated. The direct, but slightly sinuous track which separates Lower Lea Farm (bottom photo, top right) from the newly ploughed fields is a Roman Road, which emphasises further the long history of human occupation and use.

During the last 50 years the story of hedges has been dominated by neglect and destruction (photo 27). Hedges have lost their usefulness, or have become expensive to maintain, because the arable area has increased, fields have been amalgamated, electric fencing has been more widely available, and farm labour has become scarce and expensive. Hedge removal has been substantial, especially in East Anglia and the southern chalklands.

The rate of hedge loss has been a matter for debate. Estimates made in the late 1960s ranged from 1,500 to 14,000 miles per annum, the sources being unsurprisingly agricultural for the former figure and conservationist for the latter. Early detailed studies showed that 45 per cent (13,600km) of Norfolk's hedges were lost between 1946 and 1970. This, however, was at the upper end of the range; the parishes chosen for the Countryside Commission's New Agricultural Landscapes (NAL) study showed losses of 9–39 per cent over the same period. Opinions differ, too, about recent rates of loss. With fewer hedges to be destroyed, one might expect a slower absolute rate of loss, but in fact two of the seven parishes of the NAL study lost hedges faster after 1972, and other local studies have demonstrated an acceleration of destruction.

Losses have not been uniform in space or time. The Norfolk study showed marked differences between soil types, ranging from 29 per cent loss in the Good-

28 Wimpole Hall and Park, Cambridgeshire. The most obvious change between 1949 (*left*) and 1977 (*right*) has been the loss of the great south avenue, one of the more spectacular casualties of Dutch elm disease. Planted in the 1720s, the elms have now been replaced by small-leaved limes, which are so far unthreatened by any plague, and now almost fashionable. Many trees and shrubs have also been lost from the hedges to the east and west of the avenue, though the field pattern remains almost unchanged. Within Wimpole Park a few conifers have been lost near the hall, but otherwise the view endures. In the sixteenth, seventeenth and eighteenth centuries, the founding and enlargement of the park brought immense change, however. The village of Wimpole, several hamlets and their fields were swept away and replaced by pasture, leaving the cultivation remains as the sign of past activity. In the distance one sees the regular, straight-sided field pattern characteristic of recent (eighteenth and nineteenth centuries) enclosure.

sand region to 72 per cent loss on the fens. At the parish scale we are close to the scale of individual land holdings, where the preferences of individual farmers determine the whole landscape. Not surprisingly, the amount of hedge removal and the date at which hedges have been removed has differed considerably between adjacent parishes on similar land.

Whilst most lost hedges have been comprehensively cut, grubbed and burned, some have been incompletely destroyed. They remain as gappy rows of shrubs along a field bank, periodically cut back or burned. Many have been mechanically trimmed, which has eventually reduced them to lines of moribund, hollow-bottomed shrubs. Others have been totally neglected, which has allowed them to grow into rampant linear scrubs: this, however, has often proved to be a prelude to grubbing. Traditional hedge laying nevertheless survives in the Midlands, where short lengths of usually roadside hedges can still be seen with their new stakes and binders.

Trees have also long been a component of ordinary farmland. They, too, were regularly recorded in Anglo-Saxon charters. They have sometimes been so numerous that their density exceeded that of the standard trees in the neighbouring coppice woods. Like the trees in wood pastures, they were commonly cropped by pollarding and shredding, so many of our older farmland trees have the charac-

teristic wide but short trunk, surmounted by small branches in a spreading crown.

During the last 40 years, trees in farmland have suffered a sharp decline. As hedges were removed, so too were the hedgerow trees. Those trees which were retained were damaged because roots were ploughed and branches were singed by stubble burning. Field drainage and lowering of the district water table has reduced the water supply, and many trees have become stag-headed. Mature trees have been felled when saleable, but have not been replaced, partly because new saplings in hedges interfere with the convenience of mechanical trimming. Then of course there has been elm disease, which has carried away more than 90 per cent of the mature elms (photo 28) . Fortunately, the last event was so catastrophic for the landscape that it stimulated an awareness of the need for new trees in farmland, and we now see a welcome and widespread planting and retention of saplings. The scale of the decline has been measured by several local studies. In the seven parishes of the NAL study the number of trees in 1983 was between 35 per cent and 90 per cent of the trees present in 1947. In Norfolk there was a reduction of 16 per cent between 1946 and 1970. In Northamptonshire, the reduction between 1947 and 1982 was 14 per cent of the 1947 figure.

Against this background, the finding by the Forestry Commission that the number of non-woodland trees in England had increased from 54 million in 1951 to 62 million in 1980 was surprising. Moreover, the size- (and thus age-) structure was apparently satisfactory, for the smaller-size classes were well represented. The paradox is resolved by a change in the pattern of non-woodland trees. Not only are there more trees in towns and villages (photo 26), but there has been a substantial planting programme along motorway verges and other main roads. Furthermore, the lost hedgerow trees on farms have sometimes been replaced by new belts and clumps, which are large enough to count as small woods, and thus are not counted as farmland trees (photo 33). In short, we still have trees outside woodlands, but they have been redistributed.

GRASSLAND

Changes in the land use within farmland have been censussed since 1866, and most recently by the annual June Census of the Ministry of Agriculture. The most striking change has been the wild fluctuations in the extent of arable cultivation: this reached a peak of 6.0 million ha in England and Wales in 1871, but fell steadily after 1875. By 1938 arable and rotational grass extended to no more than 3.6 million ha, but since then it has increased to 5.7 million ha in 1968, then remained steady, despite the progressive loss of rural land to development. Grassland is at the other end of the see-saw. As the arable area decreased after 1875, so the grassland area rose. It remained high in the 1930s, but has since declined to well below its mid-nineteenth century extent. Recent studies by Robin Fuller enable us to show with more precision than hitherto just how great this change has been in the lowlands.

In 1932, the First Land Utilisation Survey found 7.8 million ha of grassland of all types in the lowlands of England and Wales. Of this, 0.6 million were described as 'ley'; 5.8 million as 'meadow'; and 1.4 million as 'heath'. These categories do not translate precisely into those of subsequent sources, but broadly

29 *opposite* Radcot, Oxfordshire. Unimproved grassland is often concentrated along the floodplains of the larger rivers. In this example below Radcot Lock on the upper Thames, an extinct branch of the river has left its mark as a slight depression containing fragmentary fen communities. The intensive prehistoric use of the main valleys is exemplified by small soil marks. Trees are rarely present within the grasslands, partly perhaps because the grass was treated as meadow rather than pasture, but they are characteristically abundant along the river bank.

the 'heath' is equivalent to rough grazings, including commons, whereas the 'meadow' covers improved and unimproved pasture as well as meadows in the strict sense. The great majority of the 'meadow' was unimproved pasture and meadow (photo 29), defined as swards dominated by indigenous species in essentially semi-natural communities, which will continue as such as long as grazing continues, and without the need for ploughing, reseeding, herbicides and artificial fertilisers. These and the rough pastures are collectively semi-natural grasslands, consisting not only of native grass species, but also of variable amounts of sedges and dicotyledonous herbs.

30 *opposite* Ford, Northumberland. Land improvement at low altitude (*c* 100m) in northern England has transformed extensive, bleak pastoral areas into bleak arable and temporary grass, relieved only by small patches of conifer plantations. Drains are inserted preparatory to cultivation, so the major environmental loss is the small patches of boggy ground.

Before 1939, there was some grassland improvement, but the total area of semi-natural grassland remained high. Wartime ploughing reduced the grassland area, and destroyed huge areas of semi-natural grassland. The pace of post-war change to 1971 was steady: the grassland area diminished little, but reseeding, drainage, and the application of fertilizers and herbicides caused an accelerating loss of semi-natural grassland, and a corresponding gain of artificial grassland. The Common Agricultural Policy then generated another bout of massive arabalisation (photos 27 and 30) at the expense of grassland, and accelerated even further the loss of semi-natural grassland. Despite the buffer against total extinction provided by common pastures, only 9 per cent of the 1932 area of semi-natural grassland survived in 1984. Excluding rough pasture, the loss of unimproved permanent pasture and meadow between 1932 and 1984 was a staggering 97 per cent of the 1932 area. Even this figure understates the catastrophic destruction of this once widespread habitat, for much of the residue has been partly improved by limited draining and application of farmyard manure. Now, only a small proportion of all lowland grassland is semi-natural, and some of that is not truly old grassland, but pasture reverted from arable about a century ago.

These changes are confirmed by direct ecological surveys of particular districts and types of grassland. In 1966 John Blackwood and Colin Tubbs made a comprehensive survey of all chalk grassland in England. They found 44,800ha distributed in 1,225 separate fragments, of which 931 were less than 20ha. When NCC repeated this survey in 1980, there was some difficulty about defining exactly what could be counted as unimproved grassland on chalk, but no doubt at all that the total area had much diminished. The final 1980 count was 22,800ha, the losses being due to ploughing (61 per cent), natural encroachment of scrub following withdrawal of grazing stock (32 per cent), afforestation (6 per cent) and urban spread (1 per cent). A study of Dorset chalk grassland put this in historical perspective: the area declined from 7,810ha in 1934 to 3,240ha in 1982, and that in a county which in 1793 may have had as much as 117,000ha.

The same findings are reached when all forms of unimproved lowland grassland are considered. In Dorset, Ronald Good sampled 963 patches of semi-natural grassland between 1931 and 1939. More than 40 years later Ann Horsfall found that only 24 per cent remained unchanged or had been subject only to natural changes, 62 per cent had been totally changed, mostly by ploughing, and the rest were considerably changed, though not totally destroyed or altered. In Surrey, the Surrey Trust for Nature Conservation found that 1,272ha of unimproved grassland had been lost between 1975 and 1985, of which 819ha had been ploughed, 128ha had developed naturally into scrub, and the rest had been lost to afforestation, housing, roads, etc. The low state to which these changes had brought semi-natural grassland is well illustrated by NCC's 1980 survey of the whole county of Somerset. In a county which one instinctively regards as pastoral and not intensively farmed, unimproved grassland covered just 10,200ha (2.9 per cent of the county area), and most of that was concentrated on the Exmoor fringes.

If one asks why particular fields have been left as semi-natural or unimproved grassland, one finds that a number of factors are involved. In most cases the sites are physically quite difficult to cultivate, occurring, for example, on steep slopes, shallow and drought-prone soils, or poorly-drained ground. These influences become more significant where they are reinforced by climate. Thus, some of the most important concentrations of semi-natural grassland occur on the upland fringes, such as the dales of Yorkshire and Durham. Some of the readily cultivable fields which survive as semi-natural grassland are owned by elderly farmers, who have maintained conservative farming practices. In other cases common land status has acted as a legal constraint preventing the adoption of more intensive farming practices: this is the case with the Pixey and Yarnton Mead and Port Meadow, near Oxford, which still have a pattern of grassland management established centuries ago. Finally, fields on the edge of expanding urban areas often survive for a time in limbo, neither intensively farmed, nor built upon.

The massive extension of cultivation has brought with it another bout of soil erosion (photo 35), especially in districts which have light or peaty soils, and which have few hedges and trees to baffle the wind. More subtly, herbicides and the cleaning of agricultural seed stocks has changed the flora of arable fields. Some arable flowers, such as poppy, re-appear as soon as the crop is left unsprayed, but others have been virtually extinguished, and have themselves become subjects of conservation management on nature reserves.

HEATHLAND

Heathlands are characteristically dominated by heather and other dwarf shrubs, and are maintained by grazing and burning. They are found on dry, acid soils throughout the lowlands, but are concentrated into certain districts, such as Breckland, southeast Dorset and the Lizard peninsula, to which they impart a distinctive landscape. Heathland in the strict sense is readily recognisable, but as a landscape entity it is not clearly distinguished from acid grasslands and valley mires: these vegetation types occur together in intricate mosaics (photo 21). Quantification of heathland changes thus inevitably includes some of the associated acid grassland.

The classic study of heathland change was of southeast Dorset by Norman Moore. He demonstrated not only the decline in the total area, but also the fragmentation of the habitat from a few large sites to many small sites, and the effects on the flora and fauna. According to figures from a more recent study, the heathland area declined from 18,200ha in 1934 to 5,700ha in 1983. The period of most rapid decline was approximately 1960–73, when 4,000ha were lost. In 1760 the area was around 40,000ha.

Dorset losses are typical of the other main lowland heathland districts, though each has its peculiar features. Unlike other areas, the Lizard heaths increased during the nineteenth century, but this was from a historically low base, brought about by extensive 'reclamation' for arable during the Napoleonic wars. Similar reversions from arable to heath took place in other districts, and actually produced a net increase in heathland in early twentieth-century Breckland. Losses since the 1920s have been substantial in all districts, and have been due mainly to the expansion of arable and extensive afforestation. In Breckland 11,800ha was afforested in 1920–67. A major estate in Breckland – Elveden – converted 1,215ha of heath to arable in 1927–52. The losses in the Lizard were also due to another phase of arable expansion, but here as elsewhere military uses (airfields) made inroads. Outside the main heathland districts, heath vegetation has been almost extinguished. In Hertfordshire, for example, the 83ha of heath in 1940 had been reduced to just 1.6ha by 1984.

Despite the losses, we still have 37,000ha of heathland in the major heathland areas of Breckland (photo 21), Sandlings, Surrey (photo 24), Hampshire, Dorset and the Lizard, and an additional 23,000ha scattered elsewhere in the lowlands. Much of this survives because it is common land (e.g., New Forest, Lakenheath), used for military training (e.g. Thetford, Bovington), or nature reserves (Arne, Thursley, Lizard). Few really large heaths remain. The exception is the New Forest, whose 14,370ha of heath, acid grassland and bog dwarfs the other survivors. Much of the surviving heathland is still covered by heather, but a significant fraction has deteriorated by means of natural changes following the loss of traditional management: of the 1,580ha surviving in the Sandlings, 400ha has been invaded by woodland, and 280ha is now covered by bracken.

WETLAND

The lowland landscape contains wetland in several forms. Under natural conditions extensive tracts of coastal plain and wide river valleys were so poorly drained and so subject to flooding, that they remained permanently moist and occupied by fen vegetation (photo 31). Fens naturally comprise a mosaic of marshes, sedges, reed beds, carr woodlands and wet grassland, interspersed with pools, lakes and drainage channels. Under these waterlogged conditions, humus oxidation is slow, so peat accumulates. Where rainfall is particularly heavy, or the ground water is poor in mineral nutrients, the fen becomes acid, and can even develop into raised bog (photo 32). Here, the peat accumulation is so great that the bog vegetation grows above the ground water table, and is sustained by rainfall alone. Such bogs take the form of a gentle dome of sodden, undecomposed bog peat, covered by a skin of living bog moss. The famous wetlands are the extensive tracts, such as the Somerset Levels and the East Anglian Fenland, but wetlands also occur as tiny moist patches in otherwise dry pastures, and in the meanders and oxbows alongside streams and rivers.

Drainage associated with agriculture has been one of the great landscape-transforming processes down the ages. In Britain many of the great wetlands had been drained long before the present century. Almost the final act in what was undoubtedly the greatest single wetland drainage project undertaken in Britain – the draining of the Fens – was the drainage of Whittlesey Mere (Cambridgeshire) in about 1850. This large, shallow lake set in fenland marshes and pastures was

31 *opposite* Ranworth Marshes, Woodbastwick, Norfolk. These lie in the core of the extensive fens along the river Bure, within which there are shallow lakes, or broads. Seemingly a last survival of the pristine environment, these wetlands are in fact the product of intense human activity. The broads were formed when pits created by medieval peat digging were flooded. The fens, too, were cut, both for peat and sedge. Sedge cutting is now much reduced, so the fens are increasingly being colonised by alder woodland. The trees grow in stripes because they follow the ridges of higher ground left by peat cutters. The river, too, has changed. Dredged and confined within a fixed channel, it is also heavily eutrophicated by waste from pleasure boats, and fertiliser runoff from surrounding arable. Horning Church is a bastion of stability, but the focus of the village has moved out of sight, a mile upstream.

32 Wedholme Flowe, Cumbria. This lowland raised mire fills a sharply defined basin surrounded by farmland. Originally it was an expanse of peat covered by mosses, sedges and small shrubs, bordered by a wet fen woodland. This pattern has been disrupted by peat digging and associated drainage. In the background are the narrow strips of domestic workings, which have been cut for centuries, and are still being cut in some instances. In the foreground is their modern successor, large-scale industrial sod-cutting for horticultural use. The arterial drains and their extension into the uncut mire have effectively punctured the hydrological system, leaving the uncut mire much drier than formerly. Despite this, traces of the natural patterns of surface pools survive (mid-right). The marginal fen woodland survives in patches (mid-left), but is now dry.

a productive fishery and an exceptionally rich wildlife habitat. Its destruction was hailed at the time as the pinnacle of technical achievement, the final step in the awesome task of draining the Fens, but today it is regarded as one of the greatest disasters to befall British nature conservation. Despite the long history of draining, however, traditional agriculture tolerated and used these wetlands as pasture, meadow, and places from which a harvest of reed thatch, wildfowl, etc. could be taken. The wetlands remained – albeit transformed – as part of a mixed agricultural system.

Modern agriculture has been particularly concerned to drain farmland and control flooding, in order to maximise the area of cultivation. There was, in fact, a significant amount of land drainage in the nineteenth century, when some 5 million ha were drained. This, however, was followed by a period of inactivity from about 1880 to 1930. In recent times, the rate of drainage increased massively. In 1941 5,500ha were drained. By 1945 the drainage rate was 12,000ha/yr, and by 1974 the rate had risen to 103,000ha/yr.

Under these conditions it is hardly surprising that wetlands have been lost on a massive scale. Bogs are more a feature of the uplands, but extensive tracts of raised bog were also found in several parts of the lowlands. They were best developed around the Solway, along the Forth valley, beside the rivers of southern

Cumbria, amongst the Fens, the Somerset levels, around the Humber, and across the Lancashire coastal plain, where they remained largely untouched until the seventeenth century. In 1840, intact raised bogs still covered 4,229ha of Lancashire, plus 75ha of bog damaged by peat cutting. Land-claim for agriculture proceeded rapidly during the later nineteenth century, leaving 1,005ha intact in 1900. Thereafter, reclamation for agriculture was much reduced, but drainage and peat cutting continued for horticulture, and woodland began to encroach on the dried-out surfaces. The area of open moss declined to 247ha in 1955 and 11ha in 1978. Only 2ha of this miserable remnant was close to its natural state, and even this was partially drained. By 1978, the 4,304ha of bog present in 1840 had become 3,297ha of arable cultivation, 435ha of drained moss and peat cuttings, 395ha of woodland, and 178ha of urban development. Here, as elsewhere in the lowlands, an important component of the landscape had been almost totally destroyed.

The destruction of other forms of wetland has been almost as complete. The Fens originally covered 280,000ha, but drainage and cultivation has reduced them to 855ha in four nature reserves. The small patches of fen in ordinary farmland have been greatly reduced by the wave of arable intensification. One example of this comes from Oxfordshire, where 20 small fens were identified by Druce in 1926: by the 1980s four were totally destroyed and eight were damaged by partial drainage and eutrophication. Even in places where they might be expected to be safe, wetlands have been reduced. In the New Forest, where drainage was first recorded in the nineteenth century, considerable fresh drainage was carried out in the valley mires in 1923–30. In 1965–85 the Forestry Commission carried out a further 35 drainage projects. Of 90 separate mires recently examined, 18 had been seriously affected.

In the formerly extensive fenlands, drainage created pastures bounded by water-filled dykes in place of the natural marshes. These grazing marshes retained at least some elements of the former wetlands, and remained important as semi-natural habitats, for example, in the Fens, Thorne and Hatfield Moors, Somerset Levels, Romney Marsh, Pevensey Levels, and Broadland. During the last 50 years, these have been subjected to a second bout of drainage, in order to replace relatively unproductive pasture with arable and leys. Romney Marsh, for example, was 81 per cent permanent grass and 9 per cent arable in 1939, but by 1944 the proportions were 55 per cent and 37 per cent respectively. By 1980, arable at 52 per cent was in the majority, and permanent grass had sunk to 35 per cent. Not only have the fields changed, but the ditches have been scoured, graded, treated with herbicides, and in terms of growth grossly overstimulated by pollutants. Similar developments in prospect at West Sedgemoor and Halvergate Marshes provoked a major outcry.

Open-water habitats are more likely to be altered than destroyed. Fenland drainage did away with Whittlesey Mere and many other shallow lakes in the eighteenth and nineteenth centuries. The Broads have been reduced by natural outgrowth of reedbeds and carr woodland. However, in the last 50 years, eutrophication has been the main agency of change in inland waters, mainly through nitrogen-enrichment from agricultural land and pleasure boats. This has been associated with substantial impoverishment of the aquatic flora since 1960 and extensive die-back of reed beds. Only small pools in pastoral catchments have escaped this process.

Watercourses have been very considerably altered as a by-product of agriculture (photo 33). The small streams have been converted into field ditches on such a wide scale, that they survive in natural courses only within woods, and in surviving patches of permanent pasture (photo 34). Larger streams and rivers often retain the bends of apparently natural courses, but these are dredged regularly, and the alterations of course, which naturally happened regularly at peak flows, are no longer permitted.

Numerous artificial lakes have been created over the centuries. The number of stretches of open-water in the lowlands must now be well above the natural level. Substantial lakes have been created, from monastic fish ponds and the hammer ponds associated with iron working, to the ornamental lakes of landscape parks, the feeder reservoirs of eighteenth-century canals, and culminating in the modern water supply impoundments, such as Rutland Water (which flooded 10 per cent of the old county of Rutland), and extensive flooded gravel workings.

33 Standlake and Stanton Harcourt, Oxfordshire. Old meanders seen as crop marks in arable fields. At this point the river Windrush has split into several streams, shortly before joining the river Thames. Those to the centre and right have been civilised into field ditches, but to the right the main river retains its natural sinuous course, albeit artificially stabilised. Modern tree planting can be seen both along the ditch bank, and in a rectangle by the ditched stream.

Smaller ponds were created in very large numbers in pastures, fields and woods to supply drinking water to farm stock; some were formed out of natural hollows (photo 34). Like the other wetlands which had a part in traditional farming, these too have been much reduced in number and quality. They have simply become redundant with the steady march of arable cultivation and the practice of keeping stock indoors. Three examples are typical of the decline. In Kimbolton the 152 ponds present in 1890 had declined to 103 by 1950, 67 by 1969, and 49 by 1980. Near Brighton, where there were 180 dewponds in 1930, there were 31 in 1977. In Hertfordshire, there were 7,007 ponds in 1882, but only 3,595 in 1978. Many of the surviving ponds are silted, polluted and filled with rubbish: 80 per cent of the Hertfordshire ponds were in a biologically poor condition by 1986.

34 *opposite* Site of Snelshall Priory, Buckinghamshire. The site of this vanished Benedictine Priory has reverted to permanent pasture on the eastern side of Whaddon Park. The ponds are almost certainly not the remnants of medieval fishponds, but post-medieval creations associated with the farm which replaced the priory. Although they may be a by-product of shallow quarrying, it appears that they were designed to be topped up by a small stream, a headwater of the river Great Ouse. This still meanders naturally across the pasture field (top). The farm vanished in the mid nineteenth century, but the ponds have been kept open, leaving a fringe of aquatic vegetation.

DERELICT LAND

Most of the foregoing sections have dealt with semi-natural habitats which were largely part of traditional agriculture. In modern times these were bound to decrease, the only question being by how much. Elsewhere, however, new wild places have been created, often on the sites of former farmland, through a process of use and abandonment. In quarries we have produced new cliff faces, scrubby grassland and often pools. Many reclaimed workings have been planted as woodland. Colliery spoil heaps have been put down to grass. In the valleys, substantial lakes and pools have been created by mining subsidence, and the extraction of sand, gravel and peat. Communications have provided opportunities for creating long, narrow strips of grassland, scrub and woodland along railway lines (photo 25) and motorways (photo 35). Likewise, canals have added to the diversity of waterbodies.

These are generally young habitats, which have not had time to develop. There are exceptions: the most famous is the Broads, now one of our richest wetland habitats, but there are also woodlands overlying Roman iron-workings which are indistinguishable from other ancient woodlands in the same area. Many of the railway margins are rich habitats, especially where they pass through cuttings, and where they have been bordered by species-rich semi-natural woodland, grassland, etc. Indeed, they are generally far richer now than their surroundings, which have become intensively farmed. The interesting question now concerns the new habitats starting their life in an impoverished landscape: can they possibly become as rich as the habitats which were created in earlier times?

ECOLOGICAL ANALYSIS OF LOWLAND LANDSCAPE CHANGE

It is easy to see the modern landscape as depressingly uniform, simplified and impoverished, and past landscapes as some rich, arcadian ideal. Against this background, the process of change is inevitably characterised as destruction. There are those who believe that the landscape of the eighteenth century was the optimum – enough order to satisfy the need for civilisation, but sufficient wild places to satisfy the growing appreciation of nature – and that progress since has all been downhill. In reality, there have been marked fluctuations. The pop-

ular vision of the lowland landscape, against which we judge the recent changes, has probably been more influenced by the agricultural depression before 1940, which gave us a great deal of pasture, much developing scrub, and woods that were for the first time in centuries unmanaged and thus deemed to be natural. The perceptions of those now alive start at a high point in landscape change.

Much of the landscape we inherited in the twentieth century was a complicated, small-scale mosaic of patches (fields, woods, heaths, etc.) with a plethora of connecting corridors (hedges, streams, roads, woodland rides, etc.). Cultivation had not been intense (in the modern sense), so when land went out of cultivation it reverted to vegetation of native species, though, for decades or more afterwards, the reverted heaths, chalk downland, etc., were usually floristically distinguishable from older habitats. Woodlands may have seemed like isolated habitat patches, but they were linked by hedges and streams providing semi-woodland conditions. Meadows stretched continuously along major flood plains, and the steep slopes were mainly down to semi-natural grasslands, running continuously along escarpments. Thus, semi-natural habitats were generally fairly common, and moreover they occurred in concentrations within which any patches which were temporarily cultivated could easily revert to semi-natural vegetation. Traditional land management contributed to the diversity by providing, for example, open patches and grasslands within woods, which were rich habitats in their own right; multiplied the variety of boundary zones; and moreover were links to grasslands outside the woods. When railways were built, their embankments and cuttings often developed a rich flora because they passed through a rich environment. In short, agriculture existed in a semi-natural environment, admittedly thin in places, but all-pervading.

35 *opposite* Farmland southeast of Blyth, Nottinghamshire. Erosion is not confined to deforested tropical hill slopes! Here a strong easterly wind blows the loamy sand into a dense plume across the A1 trunk road at Blyth Low Hill. The bare character of the arable fields, with fragmentary hedges and no trees, provides no means of arresting the blow. The straight-sided fields date from parliamentary enclosure, and the scatter of small secondary woods may have been planted at the same time. The verges of the dual A1 road are young habitats, which will remain free of cultivation. They are unlikely to develop into rich grasslands, however, because there are no suitable seed sources in the vicinity.

Even allowing for an unduly favourable view of the past, there is no doubt that the changes of the last 50 years have been profoundly damaging. They may not quite amount to wholesale destruction of the lowland landscape, but they have certainly impoverished its natural components on a massive scale. These changes can be summarised in the following terms:

1. Massive and general reduction in semi-natural habitats. In all the habitats examined, the losses have been on a substantial scale, even as high as 99 per cent for lowland raised mires, and 97 per cent for the once extensive semi-natural grassland. In the lowlands, these losses have been almost wholly due to agriculture. The effects of forestry and urbanisation have been small by comparison, and confined to particular areas and habitats.

2. Degradation of much of the remaining semi-natural habitats. Fens, grasslands, woods have all been impoverished by the withdrawal of traditional management. Waterbodies have become eutrophicated and polluted by the outwash of modern agriculture. In addition to direct degradation, an increasing proportion of the remaining semi-natural habitats are of recent origin in an impoverished environment, and thus they are likely to remain species-poor.

3. With the loss of semi-natural habitats and the degradation of the survivors, there has been a general loss of historic features. Even so, an amazingly high proportion of the features in the modern landscape originate in medieval and earlier antiquity.

4. It is not just the semi-natural habitats that have been changed, but the whole lowland environment appears to have been eutrophicated and impoverished. The use of fertilisers and herbicides has stimulated the growth of fast-growing weed species (nettle, cow parsley, hog weed, cocksfoot) at the expense of the greater variety of slower-growing herbs in our hedgerows and roadside verges. A much higher proportion of lowland soils than hitherto must be not only *drained*, but also *alkaline*.

5. The definition of patches in the landscape mosaic has become clearer. In the past, patches were likely to be bordered by another form of semi-natural hab-

itat, but few woodlands now grade through scrub into grassland. Grasslands are more often separated from fenland. Streams and rivers are less often bordered by patches of marshland which grade further into meadow. They are more often cut into sharply defined channels bordered by the straight edge of cultivation. We now have a high-contrast landscape.

6. Semi-natural habitats have been fragmented, and the survivors have become more isolated from each other. This isolation is due both to the actual increased separation of habitat patches, and the loss of linking habitats in the matrix. Island biogeographic theory predicts that isolation will lead to long-term impoverishment of the surviving habitats, because the inevitable local extinctions will no longer be made good by recolonisation from neighbouring patches, or from the matrix.
7. Part of this increased isolation is due to a general loss of corridors. Streams have become narrow ditches. Hedges have been so reduced that the interconnections in the network are no longer complete. Woodland rides have become overgrown.
8. New habitats have been created in old gravel pits, motorway verges, field corner tree planting, etc. However, these are less likely to grow into rich habitats than the new habitats of the eighteenth and nineteenth centuries, because they have originated in an impoverished environment, and are more isolated from rich wildlife patches.

Behind all this is a general increase in the designed landscape. Although we have long since got used to, for example, field systems which were planned on a surveyor's drawing board before they were executed on the ground, we managed to retain a great deal of the landscape which was created by individuals, who moulded their own land to their own needs using local customs and practices. Recently, however, not only have these local, unplanned features much diminished, but also agriculture and forestry have increasingly been shaped by central directives and advice. From ditch design to motorway seed mixes and the choice of species in forestry, we have increasingly conformed to national standards. What we need is a re-assertion of local character.

FURTHER READING

E. Duffy (ed.), *Grassland Ecology and Wildlife Management*, London, 1974.

R.T.T. Forman and M. Godron, *Landscape Ecology*, New York, 1986.

H. Godwin, *The Archives of the Peat Bogs*, Cambridge, 1981.

W.G. Hoskins, *The Making of the English Landscape*, London, 1955.

G.F. Peterken, *Woodland Conservation and Management*, London, 1981.

E. Pollard, M.D. Hooper and N.W. Moore, *Hedges*, London, 1974.

O. Rackham, *The History of the Countryside*, London, 1986.

O. Rackham, *Trees and Woodland in the British Landscape*, London, 1976.

3 Coastal change

England, Wales and Scotland have at least 6,000 miles or almost 10,000km of coastline. The coast is not only potentially hazardous to humans, but it is also an environment subject to continuous change. In the popular mind coastal erosion is more prominent than coastal accretion, but in fact less land has been lost than has been gained over the last century. The gains have mainly been through reclamation of mudflats and salt marshes in the estuaries, whereas the erosional losses are mostly sustained by the open coast. In either case we are dealing with an environment that is permanently and inherently dynamic, sometimes in a gradual way but often through sudden jumps and reversals after years or decades of relative tranquillity. Coastal landforms, perhaps more than most landforms, have a life not unlike that of a soldier in wartime: long periods of dull monotony interrupted by brief intervals of frightening upheaval. It is the relationship between the fluctuating life style of the coastline and the human response to it that we illustrate in this chapter.

Storms are the main instrument of rapid coastal change, causing a sudden loss of land along cliffs, beaches and sand dunes following the attack of storm waves, or the flooding of low-lying coastal land during storm surges. Such events obviously have the potential to destroy human life and property, but in what circumstances do these normal coastal processes become 'hazards'? Should we regard the hazard as a natural disaster, an act of God, or should we more properly see it as an act of Society? These questions have been hotly debated in connection with much more severe environmental hazards than any experienced today in Britain; for example, floods in Bangladesh, droughts in the Sahel and earthquakes in Armenia – events that devastate whole regions. Yet coastal hazards in Britain have the same potential to kill and destroy, though not on quite the same scale. In February 1953, 307 people drowned in coastal floods on the east coast, while in the Netherlands the death toll approached 1,800. To understand society's response or lack of response to events such as these, we need to consider first the nature of the coastal processes themselves. Secondly, we must understand how individuals and institutions react to the way they perceive these processes, which in turn depends partly on their attitudes, values and technology.

ADVANCING COASTS AND RETREATING COASTS

If our focus is on the changing coastline and on the rate and frequency of change, then the distinction between advancing coasts where accretion predominates and retreating coasts where erosion occurs is fundamental. However, both tendencies represent an average trend which may be interrupted by local or short-term reversals, and moreover a trend that is accomplished not steadily but in jumps. The

Factors affecting the long-term development of coasts.

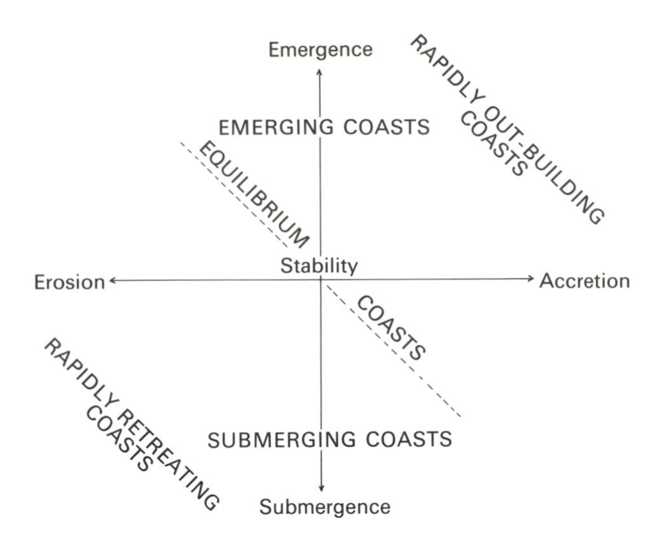

short-term and local circumstances governing change are many and varied, and are often human influenced, but in the long term advance or retreat occurs as a response by the coastline to two main factors. These are relative change in sea level, and the balance in sediment budget between coastal erosion and coastal accretion (see diagram above). A dynamic equilibrium, in other words a stable coastline, will only be found where neither force can exert a dominant influence.

The level of the sea was until recently regarded as being quite outside human influence. Sea levels will change because of the waxing and waning of ice sheets (glacio-eustasy), and because of the rise and fall of the land in response to tectonic forces or the delayed recovery of the continents from ice-sheet loading during glacial periods (glacio-isostasy). It was the glacio-eustatic process that resulted in the rapid post-glacial transgression of the sea, but in the last 4,000 years things have settled down – the eustatic sea level has fluctuated but has shown little if any long-term trend. This lack of recent trend has allowed the second set of factors – tectonics and glacio-isostasy – to have an overriding influence in many areas. Thus Scotland continues to rebound from its glacial ice sheet depression, with the isostatic rise in the land outpacing the eustatic rise in the sea for most of the last 6,000 years. As a result former shorelines have been elevated as raised beaches, often 4–12m above present sea levels (photo 36).[1]

The east coast of Scotland is currently rising by about 5cm per hundred years (for mainly isostatic reasons), while the west coast is rising in places at an even greater rate. Southern England is sinking at approximately 10cm per hundred years (for mainly tectonic reasons), reaching a maximum of over 20cm in the Thames area. In sheltered inlets like Poole Harbour, the Wash and the Thames estuary, sedimentation rates have more than matched the long-term rate of submergence, but exposed parts of the coast have been in retreat for thousands of years (photo 37).[2]

These gradual changes in land level are now taking place in the context of a renewed eustatic rise in sea level. The extent of this rise is difficult to measure, but it probably amounts to 10–15cm in the last hundred years. Already, therefore, it represents a transgression almost as substantial as the four others that have occurred on some coasts during the last 4,000 years. The two most recent trans-

36 Coastlines of emergence, Iona, Inner Hebrides, showing the eastern shores of the island, looking southwards. The abbey of Iona is built on a raised shoreline 21m above present sea level. This shoreline is late-Glacial in age, having been elevated by isostatic uplift. There are further raised beaches nearer to present sea level which date from the post-Glacial period around 6,000 years ago. This part of Scotland continues today to be gradually uplifted by isostasy at a rate of about 10cm per hundred years.

37 Retreating coastlines, Overstrand, Norfolk. The cliffs are formed of boulder clay, and are in rapid retreat through rotational slips, gullying and mudflows. From the evidence of old maps an average rate of 0.9m of loss per year has been calculated. The process is intermittent, and is often triggered by heavy rainfall or undercutting during storms. A minor storm surge in 1971 (1m rise in sea level, 3m wave height) produced overnight cliff falls at Overstrand which amounted to at least 11 per cent of annual erosion. In 1953 the storm surge approached 2m with waves 6m high, and conditions in January 1978 were almost as extreme. In between major falls sections of cliff can be virtually stable for periods of several years.

Regional subsidence in Norfolk is estimated at about 10cm per hundred years, and this undoubtedly helps to accelerate the erosion rate. Similar boulder clay cliffs on the other side of the North Sea at Gotland in Sweden, where isostatic uplift is occurring, are retreating only 0.5m per hundred years, about 200 times more slowly. Clay from these Norfolk cliffs is swept by tidal currents both westwards and southwards, to contribute to mudflat and salt marsh accretion on the North Norfolk Coast and in Essex.

38 Landscape of marine transgression, Methwold Fen, near Southery, Cambridgeshire. Today the nearest tidal river is 10km away and the Wash is 35km to the north. However, twice in the past 4,000 years tidal water has penetrated into this area. The first marine transgression started around 3000 BC and reached its maximum around 2250 BC. By the end of this phase the whole of Fenland had been turned into a huge salt marsh, flooded by high tides and intersected by an intricate creek network. The tidal flows lined each creek with silt, and today this shows up as a pale soil compared to the black peat of intervening areas. A second and less widespread transgression, affecting only the main river channels, occurred in Romano-British times between about 300 BC and the 2nd century AD. Following drainage for agriculture in the eighteenth century the overlying black peat has wasted away, revealing these salt marsh creek systems (although at ground level they are virtually invisible). The land here is at present below 1m OD, and therefore it would again revert to a saltmarsh if the tidal water from the Great Ouse was not kept at bay by levees along the river banks.

39 *opposite* Oil refineries at Coryton, and (in the background) Canvey Island in the Lower Thames estuary. Most of the reclamation of mud flats and salt marsh for industrial and urban development has occurred since 1920. The level reached by high tides has risen in this area by about 35cm in the last 100 years so a continual raising of sea defences will be necessary to protect such developments. On 1 February 1953 there was a 2m storm surge, and the sea wall around Canvey Island was breached, submerging the island. Fifty-eight people drowned and the entire population of 11,000 was evacuated.

gressions were in Romano-British times around AD 200 and in the medieval period. Each left its trace in sheets of marine sands, silts and clays in the lowlying Lancashire plain, Somerset levels and Fenland basin, to be covered by peats laid down in subsequent regressive phases (photo 38).[3]

Opinions differ concerning the causes of past episodes of transgression, but most authorities agree that global warming since about 1850 is responsible for the present rise. Slightly higher sea temperatures cause a thermal expansion of the sea, and there is also some additional melting of glacier ice. The continuing human impact on the atmosphere, particularly the increasing concentration of 'greenhouse gases', mean that an accelerating rate of global warming and consequent sea level rise can now be expected. Rises in the next 40 or 50 years of between 15 and 95cm have been confidently predicted. Despite uncertainties, the transgression of the sea that is underway is likely to represent a more sudden and a more substantial change in sea level than any that has been experienced since the early post-glacial. It will increasingly override other controls over processes, even in areas of marked isostatic or tectonic uplift, or where previously accretion has been dominant.[4]

Around the southern part of the North Sea, for example, there are protected environments where the growth of salt marshes has so far been able to keep pace with the gradual sinking of the land. This balance is likely to be disturbed, and in places like the Wash, Thames estuary, and the coasts of Denmark and Holland there are very substantial implications for existing sea defences, future reclamation and overall planning policy. In the Thames there has been enormous industrial and urban development on land that is now well below the high tide mark (photo 39).[5]

SEDIMENT BUDGETS AND SEA LEVEL

The second factor determining coastal advance or retreat is the amount of sediment supplied to a stretch of coastline relative to the processes that cause its removal. This balance of forces determines the height and width of beaches, which dissipate wave energy and also provide some direct protection for the base of cliffs. Unless they have very low slope angles, cliffs are slopes subject to the normal processes of rock weathering, soil creep, slopewash, and mass failure, leading to a flow of debris which accumulates at the foot of the slope. Where such slopes face the sea they also have marine processes operating at their base, at sea level.

Where the debris arriving at the base of the cliff is not removed by the sea, as in sheltered bays and inlets, then the slope stabilises and the coastline settles down along the line of a beach. However, where storm waves and tidal currents can remove the debris, which usually happens on open coasts and/or with soft rocks, then the land–sea slope always remains unstable. The slope becomes a cliff, it is undermined and collapses, and so the whole coastline begins to retreat inland in a vain search for stability. This recurrent process has continued for thousands of years along most of the exposed south and east coasts of England (photo 40).[6]

In this battle between the rate of sediment delivery at the base of the cliff and the rate of its removal, an important influence is the width of the beach. In the last hundred years human activities have had a far-reaching effect on beaches, by cutting off sediment supplies through sea defences or by obstructing

40 Uncontrollable erosion, part of Black Ven in Dorset, the most spectacular retreating cliffline in Britain. Interbedded clays, shales, marls, thin limestones and sands are dipping seawards, and are liable to slip at any plane where water is held up. The average rate of retreat is difficult to assess because the process is episodic. An eye-witness account comes from someone walking on the beach in February 1958: 'A great mass of Upper Greensand, complete with gorse bushes and trees, was moving down from the topmost terrace. There followed a sort of explosion in the cliff and a mass of Lias clay was hurled out. A river of liquid clay began to descend over the lower cliff on to the shore.' Any material that arrives at the base of the cliff is soon swept eastwards towards Chesil Beach, so the beach offers little protection. However, even if they were protected at their base these cliffs would continue to recede for hundreds of years towards a more stable slope angle. It is not surprising that the owners of houses on the cliff top at Charmouth have great problems insuring their property.

41 Longshore drift, Hastings, Sussex. At East Hill, Hastings, the beaches in front of the sandstone and clay cliffs have been protected by groynes since the mid nineteenth century. These structures have starved beaches to the east (right), resulting in an accelerated erosion of the cliffs. In 1850 there was an unbroken shingle beach from Beachy Head to Dungeness Point, but groyne systems are now continuous from Eastbourne to Pevensey, Bexhill and Hastings, and these have disrupted the east-going longshore drift. Downdrift of Hastings (i.e. to the right of the photograph) cliff recession is now becoming dramatic, reaching 4m per year at Fairlight where fifty houses are at risk. In 1988 Rother District Council approved a plan to construct a £2 million sea defence structure, to which the Ministry of Agriculture will contribute 70 per cent. The problem is entirely created by the Hastings Borough sea defences and by the failure of Rother planners to prevent building at Fairlight.

the longshore drift of sand and shingle through the construction of breakwaters and groynes. These obstructions may encourage the local accumulation of material, but if successful they also starve beaches in the downdrift zone, thus exposing the cliffs there to fiercer wave attack (photo 41).[7]

An estimated 55 per cent of the south coast in Kent, Sussex and Hampshire now has structural defences of some kind. In almost all cases this kind of engineering intervention robs Peter to pay Paul, an effect known in economics as a 'negative externality'. An externality exists where one person's activity affects another person but without any market transaction between them. Destructive longshore effects resulting from local attempts to stabilise shorelines are not intrinsically different from excessive use of farm fertiliser (see chapter 2), or the acid rain that spreads downwind of electricity-generating stations (see chapter 7). However, because coastal externalities are long-term and subtle they are particularly difficult to measure in economic terms. In the past they have seldom been considered at all, so that planners are not provided with an accurate overall assessment of the costs and benefits of coastal defences.

All past assessments of externalities will in any case need to be revised. The sea level rise that is now under way will exacerbate existing erosional trends, and may initiate erosion along currently stable coastlines by permitting wave action at levels above the zone protected by beaches. Where rocks are hard, as in the west of England, parts of Wales and most of Scotland, this may not have any noticeable impact. Erosion rates are very slow and will remain so (photo 42).[8] The most affected coasts will be those with soft cliffs, sand dunes or salt marshes, which include all of southern and eastern England. The vulnerability

42 Resistant coasts: the Mull of Oa, Islay, Inner Hebrides. The cliffs of Dun Athad are of hard quartzite rocks, so that despite the exposure of this southern coast of Islay the rate of marine abrasion cannot keep pace with isostatic uplift. The result is a wide, sloping boulder beach which fully protects an almost stable line of cliffs, now partially vegetated. Along this coastline even one metre of sea level rise would scarcely be noticed.

to flooding of low-lying areas will increase, and this may prove hazardous in cases where salt marshes have been reclaimed for agriculture or even urban development by the construction of sea walls. As the Anglian Water Authority admitted in a 1988 report, 'presently we build in allowance for a 1 millimetre rise in sea level a year for the estimated life of the defence. It would be prudent to increase this allowance.'

STORM SURGES AND THE THAMES BARRIER

To improve engineering structures is a reasonable and probably an inevitable response in the case of major coastal cities. In London it began to be realised after the 1953 sea floods that an extremely hazardous situation was developing.

The London Basin is subject to a long-term subsidence, and this has been accelerated by the extraction of ground water from beneath the city. After 1850 the sinking of London was accompanied by the eustatic sea level rise. Relative sea level at Tower Bridge in the period 1934–66 was rising by 43cm per hundred years, and the tidal range was also increasing by about 30cm per century.

Predicted high tides can be exceeded by at least 2.3m during storm surges, when prolonged northerly gales and low pressure in the North Sea heap up sea levels on the east coast, especially in funnel-shaped estuaries like the Humber, Wash and Thames. The biggest surge in recent times was in January 1953, which exceeded 2m at Immingham, Southend and Dover. The January 1978 surge was less severe except in the Wash, but in areas of subsidence it posed an even greater threat.

In London it was calculated that by the 1970s an area of 116sq km between Dartford and Teddington had become vulnerable to storm surge flooding. Within

43 Thames Barrier, looking towards Woolwich. During threatened storm surges, gates can be raised from the river bed to block the flood. Upstream of the barrier London is now protected from a sea level 1m higher than the storm surge of 2 February 1978, which is the biggest tide yet experienced. Downriver the effect of the barrier is to amplify sea levels during storm surges, necessitating additional flood wall protection along 23km of the lower estuary as far as Southend in the north and Dartford in the south.

this area live about 0.75 million people (1.5 million are present during working hours), and there are 15 power stations, 56 telecommunications centres, 70 underground railway stations, and several major hospitals and Government offices. Local flooding during a storm surge in 1928 drowned 14 people in London basements, and left 4,000 temporarily homeless. In the 1930s the riverside embankments were raised, but to a level that would have been exceeded in 1953 if the maximum storm surge in the Thames had not arrived three hours before high tide. There were disasters in the lower estuary, for example at Canvey Island and the Isle of Sheppey, but London escaped flooding. Further increases in embankment height narrowly averted another disaster in the storm surge of 1978. Only in 1983 was relative security assured with the construction of the Thames Barrier (photo 43).[9]

This remarkable construction cost over £800 million, and was designed to contain a tidal flood with a 1,000-year return period. In fact the level against which London is now protected is only 7.2m, which is 1.0m above the actual level reached by the major storm surge of 1978. Some global models now being seriously considered predict at least a metre of eustatic sea level rise within the next 50 years, because of the greenhouse effect. If this should come about then the Thames Barrier will be obsolete, and within a generation central London will again be threatened by impending catastrophe.

THE RESPONSE TO COASTAL HAZARDS

Where a hazard is less extreme than the flooding of a city, or is less well understood, then the response of a society is not necessarily to defend the status quo at all costs. In pre-industrial times coastal populations adapted to coastal hazards by recognising there were limits to their capacity to prevent nature from taking its course. In Fenland, for example, all settlement was formerly located on solid ground that was thought to be relatively safe from floods, and the economy was based on grazing, fishing and fowling as much as arable. Away from the fen edge and fen islands like the Isle of Ely, the silty beds of former tidal channels (roddons) provided some sort of protection. After drainage these roddons became slightly elevated above the surrounding fen, because of the wasting of the peat (photo 44).[10]

44 *opposite* Adaptation to hazard, Fenland. The winding band of silt, or roddon, is a former tidal channel of the River Lark at Littleport, Cambridgeshire. Tidal silt deposited in the river channel is now slightly elevated by the wasting of surrounding peat, following drainage. The roddon has provided secure foundations and dry sites for farms and cottages. Littleport itself, in the foreground, is situated on a fen clay island well above sea level. The straight river channel is an artificial channel of the Great Ouse, cut in the medieval period to help in the drainage of Fenland.

Despite their awareness of risk and their adaptations in land use and settlement, coastal populations did not always avoid disaster. The medieval period in particular was a time of increasing vulnerability to coastal erosion and coastal flooding, especially around the North Sea. This has been attributed to climatic deterioration, storminess and a rising sea level, but the existence of large numbers of people living in disaster-prone areas also requires explanation. It is a situation with interesting parallels to the modern Third World, where landless peasants often have little choice but to reclaim and settle marginal and potentially hazardous areas where their vulnerability to disaster is inevitably increased. Marshland, the region of Norfolk adjacent to the Wash, was flooded twelve times between 1250 and 1350. On one such occasion, in 1337, the people of Tilney reported that their sea bank was broken 'by the raging of the sea' in five places, so that the village was inundated:

> The lands, meadows and pastures were continually drowned for the space of seven days, by which means their winter corn, then sowed upon the ground, was destroyed, as also much of the corn and hay in their barns, with one hundred muttons and sixty ewes . . .
>
> (W. Dugdale, *History of Imbanking and Drayning*, 1662)[11]

Elsewhere around the North Sea the reported death tolls from storm surges were counted in hundreds of thousands. It was at this time that the Zuider Zee in Holland was formed by sea floods, and perhaps also the Jadebusen Bay in north-west Germany and the Broads of Norfolk. Huge tracts of land were eroded in western

Denmark, and in eastern England the great ports of Dunwich (Suffolk) and Ravenspur (Humberside) were lost. It was a time of blowing sand, as once stable dune systems broke up and began to move inland (photo 45).[12]

How could people explain such natural disasters? The conventional view was fatalistic. Before the eighteenth century nature was viewed as an artefact whose inner workings were beyond human rationality. Disasters, like moral evil, were expressions of some divine purpose, and were often seen as lessons and warnings to society, perhaps even as punishment in extreme cases. Not until the Enlightenment was this view of natural hazards seriously challenged. The earthquake that destroyed Lisbon in 1755 provided an opportunity for a debate regarding the

true cause of such phenomena, whether physical, moral or religious. The scale of catastrophe in Lisbon even raised doubts in the minds of some regarding the competence of the Creator. On the one hand, in a perfect created world, 'All nature is but art, unknown to thee; / All chance, direction which thou canst not see; / All discord, harmony not understood; / A partial evil, universal good' (Alexander Pope, *Essay on Man*). On the other hand, as Voltaire pointed out, to attribute a sense of benign purpose to all events in nature was absurd. Were the tides created, as some had claimed, in order that ships could go in and out of ports more easily?

TECHNOCENTRIC AND ECOCENTRIC ATTITUDES

The Enlightenment, the growth of science, and Europe's exploration of the New World and beyond, destroyed for ever the view of nature as a work of art fulfilling some divine purpose. In western thought the notion that natural disasters were lessons, warnings or punishments was replaced by a view which saw hurricanes, earthquakes or floods as being explained by processes acting independently of human existence. We should look to science to explain the causes of such phenomena and their effects. The study of hazards became dominated by geologists, hydrologists and engineers, and later by social scientists.

The instinct of these investigators, and indeed their assigned role in society, was to use their growing understanding of nature's workings as a basis for increasing human control over nature. A 'technocentric' philosophy become dominant, emphasising rationality, an 'objective' appraisal of means to achieve given goals, and policy control by a professional elite. The approach is typified by the Royal Commission on Coastal Erosion established in 1906, which was required to evaluate the extent of sea encroachment and the most appropriate means for its prevention. The Commissioners recommended in their report that a central authority should take charge of British coasts, in order to coordinate all scientific and engineering efforts. In the UK such a body was not established, but in the USA the Corps of Engineers and the National Park Service began to play a central role, promoting coastal protection at all costs and by the best engineering means available. The objective of control over nature was never questioned.

It is only quite recently that an alternative 'ecocentric' view of nature and society has again become prominent. This alternative world view can be traced back to romantic philosophies current in mid-nineteenth century America and elsewhere, when a respect for nature was seen as the only sound basis for democracy, religion and art. Social and spiritual harmony was associated with self-sufficient communities (Peter Kropotkin), or with Garden Cities (Ebenezer Howard).

Ecocentrism of this kind was easily rejected as utopian. In the UK the movement seemed to be overtaken by the establishment of Town and Country Planning measures in the 1940s, but in different form it re-emerged in the 1960s as the ideology of a new ecology movement. With its new 'global ecosystem' context, the movement is characterised by a reverence for natural processes, a concern for conservation and sustainability, and a distrust for technocentric solutions that is linked to a distrust of the centralised state. The influence of ecocentric thinking

45 *opposite* Migrating sand dunes, Kenfig Burrows, between Margam and Porthcawl, South Wales. The ruins of Kenfig Castle are visible on the seaward side of the modern railway line. Buried beneath the dunes in this photograph is not only the castle and its moat but also a church and small town. Kenfig was overwhelmed by blown sand some time after AD 1300, and historical accounts report the same problem elsewhere on the South Wales coast. There is a possible connection with overgrazing, but there is also evidence of coast erosion, flooding and storms, probably linked to climatic change and sea level rise. Beneath sand dunes at Margam is the site of the abandoned Hermitage of Theodoric. When excavated the buildings revealed thin plastered clay walls supported only by the sand, suggesting quite sudden burial by the dunes. Today the dunes are stable apart from a few blowouts, and are covered in dense thickets of sallow (*Salix repens*).

46 *opposite, top* Unregulated coasts, Holderness, looking south towards Tunstall. The boulder clay cliffs at this point average over 20m high, but like elsewhere in Holderness the historical record is one of continuous retreat. Nine hundred years ago the farmland in the foreground belonged to the village of Monkwike, which was recorded in Domesday Book but had all disappeared by the middle of the last century. The old road from Tunstall to Monkwike can be seen ending at the cliff edge.

Between 1852 and 1952 the rate of retreat along Holderness averaged 1.2m per year, and one estimate suggests that altogether about 215km² has gone since Roman times. Material from this coast is transported by currents south towards the Humber, Wash and perhaps beyond, where it contributes to salt marsh accretion. Meanwhile the Holderness Coast Protection Project hopes to raise £1 million for a pilot project for an offshore barrier, the total cost of which may exceed £200 million and will require at least a thousand million tons of rock debris. The organisers hope that the government and the EC will share the cost.

is growing in many western democracies, in Britain through informal lobbying rather than formal representation (see chapter 7).

In relation to natural hazards ecocentric approaches have had little effect so far in this country, but in the USA there has been a major reassessment. The losing battle of the American engineers against erosion, for example, absorbed increasing amounts of public money from the 1930s to the 1970s, but was accompanied by a rising level of property damage from storm events. Along the retreating barrier coastline between Florida and Maine the policy is now to adapt to the secular trend of erosion, which occurs intermittently and usually during major storms, rather than attempt in between storms to reverse the trend. In the states of North Carolina, Maine and Massachusetts the construction of hard sea-defence structures is now illegal, but beach replenishment is permitted. Only the major cities such as New Orleans, Miami and Manhattan will persevere with traditional structures, despite the mounting costs. If we enter a phase of more rapid sea level rise and perhaps more hurricanes, this will pose a challenge to the policy of allowing the coastline to establish its own equilibrium. However, in a period of budgetary restraint in Washington, ecocentrism may provide the federal government with a welcome legitimation for further reductions in public expenditure on coastal defence.[13]

LETTING NATURE TAKE ITS COURSE

The two largest stretches of unregulated coastline of east-coast England are the Holderness cliffs of Humberside and the North Norfolk coastline of barrier beaches, sand dunes and salt marshes. Holderness is a 61km stretch of soft boulder clay cliffs, retreating in the century up to 1950 at an average rate of 1.3m per year. Beaches are absent along most of the coastline, and full protection could only be achieved at prohibitive cost (photo 46).[14]

In any case, the consequences of stabilising Holderness might be quite negative for other areas. There is probably a connection between erosion there and accretion further south, at least as far as North Norfolk. The 70,000 tons of mainly fine sediment removed each year from Holderness cannot all be deposited on the bed of the North Sea or in the Humber and Wash estuaries. Salt marsh accretion in North Norfolk has managed for at least 2,000 years to match a local rate of subsidence estimated at 8cm per hundred years, despite the marshes having few local sources of sediment.

However, sand and shingle are less readily available to the North Norfolk coast than mud, and to accommodate the gradual sea level rise the seaward beaches and dunes have migrated landwards, as on the east coast of USA. The process of adjustment is accomplished not steadily but in jumps. At Scolt Head Island major storm surges in 1938, 1953 and 1978 pushed back the island, as storm waves cut back the high dunes and washed over the low dunes (photo 47).[15] In the 1978 storm an average width of 20m was removed from dunes, but most of this material later moved inshore again and replenished the beaches. Longshore drift has also healed some of the breakthroughs, and new dunes have formed. In between major storm surges the coastline seems to take one step forward, only to be pushed two steps backwards by the next major storm.

47 Unregulated coasts, Scolt Head Island. A photograph taken a few days after the January 1978 storm surge, which cut back 20m of shoreline and overtopped and destroyed the belt of low dunes. The marshes on the left of the picture were flooded to a depth of 2.5m but were undamaged. Beds of marsh mud from a former salt marsh can be seen outcropping on the seaward beach. Most of the eroded beach and dune material was deposited on top of former marshes as washover aprons, well above normal high spring tide level. In the decade since this storm these washovers have been the focus for new dune development. In this way the island's coastline adjusts to the constraints of a negligible supply of new shingle and a gradually rising sea level. To intervene successfully in this process would require gigantic engineering structures: anything less would be futile. From the conservation point of view intervention is, in any case, undesirable.

In this way natural coastlines adjust to their restricted supply of beach sediment and to the stress of rising sea level. At Holderness there is some pressure by landowners for intervention in the process of erosion, but most of the North Norfolk coast is owned or controlled by conservation bodies, and their interventions have been small in scale. An attempt was made after the 1953 breakthrough at Scolt to stabilise the line of dunes, but as well as being futile such a policy is undesirable. New habitats are created as the island adjusts to natural processes, whereas fixed shorelines will lead to an ecological succession that eventually will eliminate its distinctive character. At Scolt Head Island nesting terns were once found only on pioneer sand dunes at the western end, but they have now colonised also the new shingle washovers created by the 1978 storm.

48 Stopping nature taking its course, Orford Ness. This large spit of shingle on the Suffolk coast has diverted the course of the River Alde, so its mouth is now 18km south of Aldeburgh, the town in the foreground. Meanwhile the supply of shingle to Orford Ness, which used to arrive from the north via the beaches at Aldeburgh, is now held in place by large-scale groynes in front of the town. The small groynes on Orford Ness itself cannot prevent a narrowing in the neck of land between the North Sea and the river, because of erosion on both sides.

Taken in 1949, before the 1953 storm surge which pushed Orford Ness landwards by 50m in places; the photograph shows attempts at replenishment of beaches by bulldozing, and shows also the construction of groyne defences near the Martello Tower. Even so this whole neck of land was overswept in 1953 and again in January 1978. J. A. Steers has remarked 'To try to protect the whole of Orford spit would be not only difficult but outrageously expensive. Moreover it would almost certainly be useless . . . As a result of the defence works at Aldeburgh and Slaugham the damage done by further storms cannot be healed by nature.'

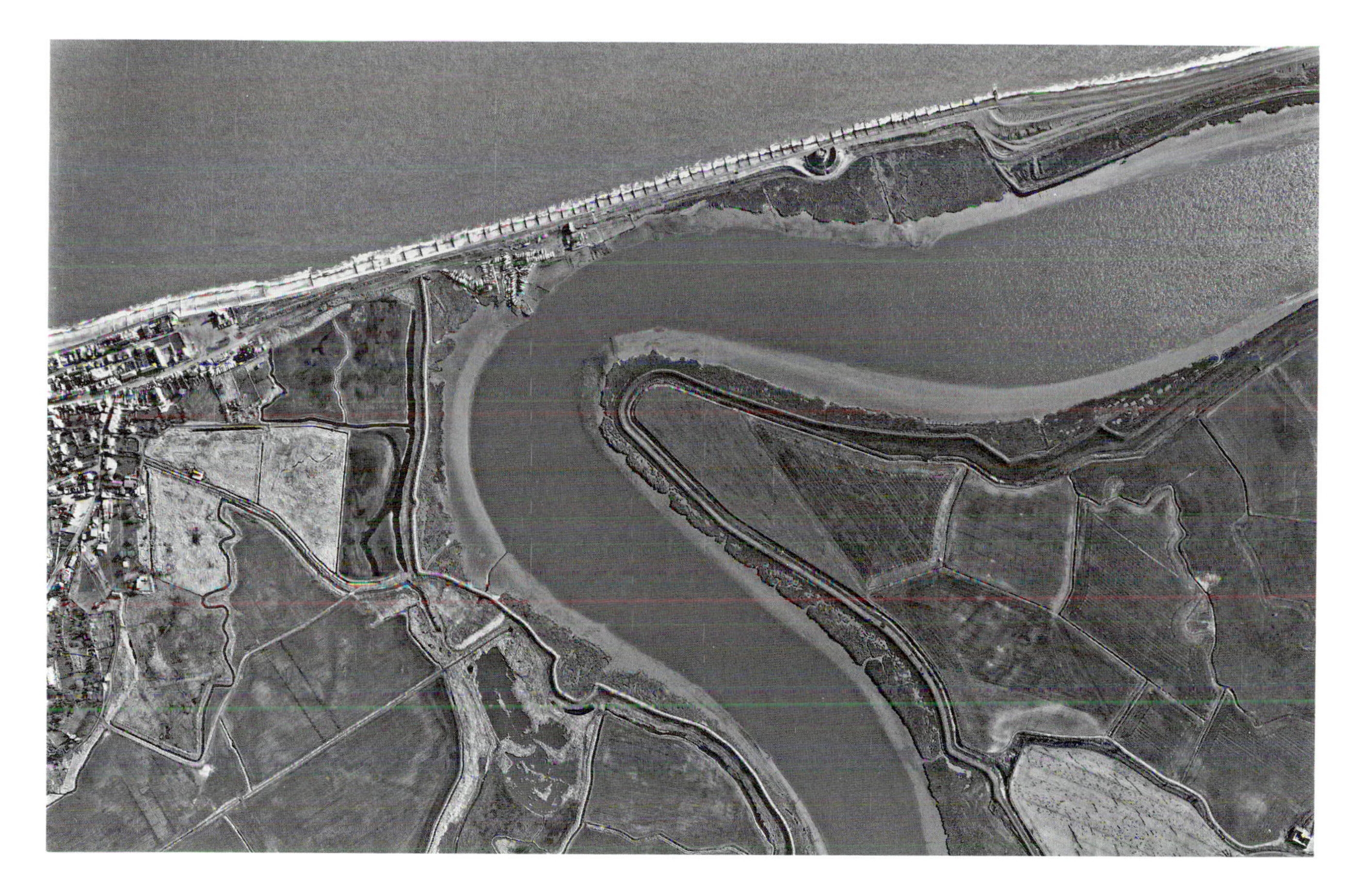

There is a price to pay for letting nature take its course. The Scolt Head Island National Nature Reserve has lost 30–40m of land since 1953 along 5km of coast, but elsewhere in North Norfolk and in the Wash new salt marshes and dunes are being formed. Since resources for conservation are scarce it is far more sensible to concentrate on acquiring these new areas rather than attempting an undesirable and probably futile protection of existing ones, by means of expensive engineering works (photos 48, 49).[16]

49 Vertical air photograph from 1983, showing Orford Ness still surviving at the Aldeburgh end, but only because of a continuing commitment to sea defences which steadily diminishes its natural character. Little more than a narrow sea wall now separates the North Sea from the Alde. In 1988 a town councillor was quoted as saying, 'Some experts say the wall will last another five years, but people who live here think that is nonsense. We can't say when it will happen, but if we got a series of gales it could be this month.' The implications of allowing the River Alde to find a new mouth near Aldeburgh would be to make Orford Ness into an island, and to allow tidal flows to penetrate further upriver, increasing the risk of sea flooding at the famous Snape Maltings concert hall.

DEFENDING THE COAST: PRIVATE GAIN AT PUBLIC EXPENSE?

Holderness and North Norfolk have avoided being 'protected' by engineering structures, except locally, because of their sparse coastal settlement and low economic value. Elsewhere it is a different story. In southeast England particularly there is more than clay cliffs and sand dunes to defend: there are harbours, housing estates, hotels and farms, not to mention the professional interests of the engineering companies, and the reputations of the public bodies, the district councils with responsibility for coastal erosion, and the recently privatised water authorities charged with protecting land from coastal flooding. All these parties tend to assume that sea defences are in the national interest, whereas in fact structural engineering usually results in private profit at public expense. The true public interest would usually be served much better by a quite different interventionist policy.

The situation has not changed fundamentally since the Royal Commission on Coast Erosion made its final report in 1911, despite much urban development along the coastline and agricultural intensification in low-lying areas protected by sea walls. In 1911 the Commissioners identified the need for more research and emphasised the advantages of coordinated action, but the financing of sea defences was another matter. In their conclusion they stated, 'we cannot see there are any grounds for the contention that sea-defence is a national service; it is true that there is serious erosion in places, but this erosion does not affect the nation at large'. They argued that the making of grants from public funds in aid of sea defence should be discouraged, on the grounds that any money so granted would protect private property not state property, and would thus represent a subsidy to private landowners from the ratepayers at large.

Until recently this principle was largely forgotten. Until 1952 there was a proliferation, not a centralisation, of the agencies for planning the coastline, and an increasingly sophisticated but expensive capacity for structural intervention. The River Boards, later called Water Authorities, gained more control after 1952, but management of the coast remains uncoordinated and piecemeal. As an example, the 45km stretch between Brighton and Hastings is divided into fourteen sections which at present are managed by nine separate bodies including three Borough Councils, three District Councils, the Newhaven and Seaford Sea Defence Commissioners, the Southern Water Authority, and (at Newhaven Harbour) British Rail (photo 50).[17]

50 Uncoordinated planning, Birling Gap, Seven Sisters coast, Sussex. Apart from local sources of shingle from cliff falls, there has been no longshore drift of beach material from the west for over a century because of harbour protection at Newhaven and groynes at Seaford. At Birling Gap, despite the presence of a small beach, the rate of recession of the cliffs averaged 0.9m per year between 1875 and 1961. On the higher and more resistant cliffs of the Seven Sisters (background) and Beachy Head, the rate averaged only 0.42m per year. Some coastguard cottages at Birling Gap dating from the late nineteenth century have been destroyed by the sea, but as late as the 1960s planning permission by Wealden District Authority was being granted for the construction of new bungalows. Meanwhile, Southern Water Authority do not intend to attempt any protection for the cliffs, now part of Seven Sisters Country Park, one of Britain's Heritage Coasts.

51 Coastal speculation, Peacehaven, Sussex. In 1915 Charles Neville, a businessman, purchased 650 acres of open downland on the cliff tops between Saltdean and Newhaven, in order to develop a new town. The name he originally chose was Anzac-on-Sea. Plots were offered for £50, £75 or £100 depending upon their nearness to the cliff top. Building started in 1918, and by the Second World War a straggling eyesore of bungalows and holiday shacks had emerged, unserviced by made-up roads, water supplies or sewers. In the Depression some unsold plots were given away as prizes in newspaper competitions, in an attempt to complete the struggling project. In the post-war period infilling and rebuilding has turned Peacehaven into a more normal south coast town, and permanent roads, sewers, etc. have been constructed by the local authority.

Below the cliffs at Peacehaven beach replenishment is now negligible because longshore drift from the west has been interrupted since 1862 by large-scale groynes at Brighton, extended in 1928 to Rottingdean and Saltdean. Downdrift of Saltdean, at Peacehaven, the rate of cliff erosion is accelerating, and Lewes District Council (which has reluctantly accepted responsibility for Peacehaven) is under pressure from residents to intervene to protect their property. The final price for Charles Neville's speculation has yet to be paid.

It goes without saying that administrative boundaries are not geomorphological ones. Many systems of sea walls stop at town boundaries, and adjacent authorities may pursue quite different policies. Moreover responsibility for the zone immediately inland of the coast is subject to town and county planning legislation that may be administered by quite separate bodies. Their policies for development of the hinterland may not at all match the policies pursued for coastal protection (photo 51).[18]

Since 1952 the Water Authorities have controlled the lion's share of the coastline. These are curious institutions, dominated by professional engineers and by agricultural interests, and having large and unregulated financial powers. They have a statutory duty to supply water, regulate rivers and prevent flooding, but no responsibility to the wider public interest. In the last 40 years a large and continuing commitment to sea defence has been favoured, in order to stop the erosion of cliffs, retain beaches and prevent sea floods.

Behind these defences, all erected mainly or wholly at public expense, housing developments occurred and farmers began to intensify their land use, particularly during the long boom in subsidised agriculture from 1940 until the mid-1980s. Aided by 75 per cent grants from Ministry of Agriculture land drainage was improved, and with guaranteed EC prices there was a widespread switch from rough grazing to arable farming. The lessons of the 1953 floods were misinterpreted or forgotten. In King's Lynn the 1978 storm surge inundated new housing estates, factories and a NHS hospital all constructed in flood-prone areas below sea level, and in a region subject to a relative sea level rise of at least 35cm in

the last hundred years. In this way an estimated 27 per cent of the coastline in England and Wales has been developed, mostly since 1920. When floods or cliff falls do occur, the actual or threatened property damage is interpreted as making necessary a further strengthening of the defences (photo 52).[19]

THE TURNING OF THE TECHNOCENTRIC TIDE?

Because of a consistent lack of vision by central government, unregulated technocentrism has led to large negative externalities and an escalating level of state subsidy for sea defences. In many instances existing defences will be overwhelmed by the predicted sea level rise during the next 50 years. As a result of this failure of the state to achieve holistic planning, coastal hazards that were once seen as acts of God inflicted on wicked society must now be regarded as acts of Technocentric Society, only kept at bay by the taxpayers, few of whom directly benefit.

An alternative, non-structural response to coastal erosion and coastal flooding is feasible, by means of beach replenishment, strict land use zoning in areas that are permanently at risk, and efficient warning systems. A turning point in this direction was the decision by central government in 1982 not to give grant aid

52 *opposite* Private profit at public expense, Wells-next-the-Sea, Norfolk. Natural salt marsh is in the foreground. In the distance is part of the sea wall built in 1758 by the Duke of Leicester as part of his drive for 'agricultural improvement'. The second sea wall (dating from 1859) extends from left to right, and borders the Wells Harbour Channel. It was breached in 1953 and again in 1978, shortly before this photograph was taken. Both sea walls originally were the outcome of private capital invested for private gain, but the maintenance costs have increasingly been taken over by the state. In 1978, when this breach occurred, the Anglian Water Authority chartered helicopters to drop hundreds of tons of chalk into the gap. This expensive exercise was the prelude to closing the breach with bulldozers, and the entire sea wall was subsequently raised and strengthened. In this case the works are of no value for the protections of lives and buildings (none are at risk), but they do enable 650 acres of arable farmland to continue in intensive wheat production.

53 Structural intervention prevented: Breydon Water (at low tide), site of the proposed Yare Barrier. Flowing across Breydon Water is the river Yare, which is joined by the river Waveney in the foreground on the right. The proposed £20 million barrier would have been located at the far end of Breydon Water, near Great Yarmouth. This structure was designed by Anglian Water Authority to prevent high spring tides and storm surges from posing a threat to 21,000ha of low-lying marshland that extends inland as far as Norwich. The consequence would have been to enable farmers to convert their land use to more profitable arable farming. Central government decided not to subsidise the scheme, now abandoned.

to the Yare Barrier Scheme in Norfolk, designed to protect 21,000ha in the lower valleys of the Yare, Bure and Waveney rivers. All this area is below river level and so is liable to saltwater flooding during major storm surges, which last occurred in 1953 and, on a small scale, in 1978 (photo 53).[20]

In this area protection from normal high tides is prevented by earth embankments along the river banks, maintained by local drainage boards at an annual cost of £350,000 in 1980 (85 per cent grant-aided). The proposal from Anglian Water Authority was to construct a £30 million tidal barrier to provide total protection from the effects of high spring tides and storm surges. An alleged 'benefit' would have been to enable farmers to change from cattle grazing on marshland to arable farming (itself state subsidised). A hidden cost would have been the loss of the characteristic Broadland landscape, as well as the distinctive flora and fauna of the dykes and pastures.

The Yare Barrier scheme aroused a large public protest because it symbolised the injustice of private profit being achieved through state subsidy, and at the cost of the wider public interest and the conservation value of a unique landscape. The scheme was also proposed at a time when EC food surpluses and the whole subsidised basis of agriculture were coming under scrutiny, and at a time of major cuts in public expenditure. It was not approved by the Department of Environment, although ironically the compromise drainage measures that were approved and funded by government later required a rescue package for the Halvergate Marshes, threatened with arable farming despite the lack of complete sea flood protection.

CONCLUSION

A side effect of the free-market philosophy of Conservative governments since 1979 has been a re-examination of the legitimacy of environmental interventions funded from public expenditure. In relation to coastal management there are some signs of a turn in the tide of opinion that once gave approval to sea defence as a matter of course. Government policy towards agriculture has had to be redefined: 'Support methods which were appropriate in the early years are no longer sustainable now that a pattern of significant excess of supply over demand has emerged' (Ministry of Agriculture, *U.K. Farming*, 1987). On the coast it became official policy in 1984 to regard 'undeveloped' sections of coastline as being not appropriate for grant aided protection.

As a result of this shift in government policy a temporary alliance has emerged in British politics between the environmentalists and the free marketeers. This latter group wishes to reduce state subsidies and interventions, in order to reinstate free-market forces. The environmentalists are in favour of the first part of this programme, particularly if it prevents large-scale technological solutions, but only as a prelude to a new era of holistic planning. In relation to coastal management, the alliance will collapse should the coastline be fully opened up to capital.

On both sides we are still a long way from a full recognition of the interdependent nature of coastal processes. The problem was diagnosed in the Royal Commission Report (1911) and restated in the Report of the National Parks Committee (1949), but holistic coastal planning has still not emerged. There is as yet no blueprint for the proper coordination of coast and hinterland planning, despite the growing tensions between a hinterland that is becoming more developed and a coastline that is likely to undergo changes more rapid than any experienced for 6,000 years. Our response to the current transgression of sea level will depend on how we interpret the lessons of the past.

FURTHER READING

H.C. Darby, *The Changing Fenland*, Cambridge, 1983.

J.D. Hansom, *Coasts*, Cambridge, 1988.

H.H. Lamb, *Climate Present, Past and Future. Vol. 2. Climatic History and the Future*, London, 1977.

E.C. Penning-Rowsell, D.J. Parker and D.M. Harding, *Floods and Drainage: British Policies for Hazard Reduction, Agricultural Improvement and Wetland Conservation*, London, 1986.

A.H. Perry, *Environmental Hazards in the British Isles*, London, 1981.

J. Pethick, *An Introduction to Coastal Geomorphology*, London, 1984.

D.T. Pugh, *Tides, Surges and Mean Sea Level*, Chichester, 1987.

J.A. Steers, *The Coastline of England and Wales*, 2nd edn, Cambridge, 1964.

M.J. Tooley, *Sea Level Changes: Northwest England during the Flandrian Stage*, Oxford, 1978.

4 Leisure and the countryside

Since the Second World War Britain has undergone a series of important social and economic transformations. Not least, there has been an enormous increase in availability of effective leisure time. Structural changes such as better pay, a shorter working week, longer holidays, widespread personal mobility in the form of the private car and now a bulging and affluent middle-aged population have worked together to produce a seemingly unstoppable demand for all kinds of recreation activities. Nowhere is recreation demand more apparent, or has the brunt of consequent pressures caused so much concern, as in the countryside. The scale of countryside recreation is extraordinary. The countryside is visited by over 80 per cent of the population every year,[1] and with public interest at this level is inevitably an important target for both private investment and public policy. Leisure has become big business and the commodification of recreation is well advanced. All the paraphernalia marketed in the wake of projected social images and lifestyles, specialist outdoor clothing and camping equipment for example, reinforce the leisure orientation of society. Of course, people generally are better educated and much more aware of available opportunities, but it is a moot point to what extent these trends actually stimulate new pressures rather than simply release pent-up or latent demands that were not previously met.

Governments recognise that increases in leisure may not necessarily have wholly positive outcomes. Leisure may be a social opportunity but at the same time it suggests potential for social and environmental problems and there is much scope and need for public sector management of countryside recreation resources. There is now a new breed of leisure professionals, staffing the Countryside Commission and in the leisure and planning departments of Local Authorities for instance, most of whose jobs did not exist before 1960. Yet relatively little countryside is publicly owned for purposes of recreation. Measures that have been taken to protect the countryside simply express a political desire and the power to extend public control over the use and development rights of riparian owners. The National Park Authorities are the most striking case in point. Taken overall, the ten Authorities actually own a mere 2.3 per cent of the area within their administrative boundaries[2] so that their real function is as a watchdog over land that is otherwise owned, and is a multi-purpose resource used variously for agriculture, forestry, water supply, defence and a whole host of other purposes besides recreation. In fact the period of explosive growth in recreation demand coincided with other large-scale changes, mainly bound up with massive expansion and development in agriculture and forestry (see chapters 1 and 2):[3] changes that have so affected landscape and habitat as to form the basis of the triangular interlocking conflict between agriculture and forestry, recreation and conservation that makes up a large part of the rationale for this book.

54 *opposite* Woburn Abbey, Bedfordshire. In 1955 this stately home became the first to be opened to the public and is now said to attract over one million visitors per year. The house, mainly seventeenth century and eighteenth century by Henry Flitcroft and Henry Holland, is set off by a 1,100ha deer park landscaped by Humphrey Repton. Attractions today include a safari park, cabin lift, boating lake, fairground rides and a parrot show.

Although the period of hitherto unimagined growth in recreation participation took off around 40 years ago, antecedents of the pressures we now see in the countryside have deep roots. The countryside commands a special niche in the psyche of the British public. In an historical sense we are more or less all of the countryside and there is intuitive appeal in resting our explanation for the popularity of countryside recreation in this simple fact. Yet Britain's population was already more than 50 per cent urban by 1851 and whatever our present day images of a past rural idyll it is doubtful whether the hardships of the pre-industrial economy permitted much opportunity for leisure or anything that we would recognise as countryside recreation. A round of feast days and very localised pleasures perhaps, but certainly not the multifarious demands placed on resources by today's affluent, and increasingly retired, post-industrial society. The countryside evolved as the preserve of an extremely restricted elite and in a very real sense the key issue in the development of countryside recreation has been the question of who owns the land.[4] The idea that landownership should not necessarily carry with it an unfettered right to determine control over and access to the resource is comparatively recent, though the extension of public policy into countryside planning and management, together with the growth in car ownership, has lowered many barriers that formerly inhibited recreational use.

In extreme cases landownership has been continuous since Norman times. To take just one example, Belvoir Castle in Leicestershire (see photo 110 in chapter 7) was built by one of William the Conqueror's followers and now, over 900 years later, the present Duke of Rutland can feel content that he still holds 6,000ha, never sold since first appropriated. But at the top of the landowning tree Rutland is dwarfed by the Duke of Buccleuch, reckoned as Britain's biggest private landowner (116,000ha, mainly in Scotland). An 1876 land return (the last and only one on record!) revealed that three-quarters of England, Wales and Scotland was in the hands of a mere 7,000 landowners. In England and Wales around 700 people owned 25 per cent of the land. The system of landownership stretching back to the Conquest and in particular the concept of freehold property rights, had determined development and controlled accessibility in the countryside for hundreds of years. Paradoxically all is not negative for although the countryside has evolved by the whim of the few, now it is *there*, a heritage in all its magnificence and diversity. Much of the medieval countryside, for example, was given over to forest (private hunting chases and deer parks) but that is why we are able to enjoy surviving examples, such as the New Forest and Staffordshire chases, today. Many of the stately homes of England, built mostly after 1600, were set off in extensive private parklands. Blenheim (1,100ha) and Woburn (1,000ha) (photo 54) are fine examples and their parks and gardens, laid out by pioneering landscape gardeners like William Kent (1685–1748) and Lancelot 'Capability' Brown (1715–1783), founded a new aesthetic tradition that has deeply affected people's perception of 'natural' countryside. Today their created landscapes are amongst the most prized and cherished and, following the effect of death duties and unaccustomed financial stringency for some members of the aristocracy, many of the best houses and gardens have been opened to the public and attract literally millions of visitors each year.

55 *opposite* Buxton, Derbyshire. Famous for its natural warm springs since Roman times the town became a mecca for fashionable society when in 1780 the Duke of Devonshire decided to divert some of his copper mining wealth into its redevelopment as a spa to rival Bath. The design centres on an ambitious complex of baths and other fine buildings adjacent to 9ha of parks and gardens bordering the river Wye. Most famous is The Crescent, clearly visible in the photograph, and behind it the Great Stables. In 1858 the latter was converted into the Devonshire Royal Hospital for rheumatic patients, the courtyard being covered over by a remarkable dome in 1879. Other important buildings included the Palace Hotel and Pavilion. The last major building was the Opera House not completed until 1903 by which time the heyday of the spas had passed.

While the countryside was for the elite the true antecedents of mass participation were the spas and seaside resorts of the seventeenth and eighteenth centuries.[5] Although their reputations hinged on the health fads of the times and the supposed curative value of their waters, the spas were always geared to a numerically small, high-class, fashionable clientele. Accessibility was vital. At first the spas were near the capital. Tunbridge Wells in Kent gained early fame. Later, as transport and communications improved places farther afield became popular. Leamington, Cheltenham and Bath (by far the largest and most successful attracting upwards of 10,000 visitors per year in its heyday) are well known. Towards the end of the era Malvern and Buxton (photo 55) were set up in competition. The spas were the model that prepared the way for seaside resorts and the promotion of forms of leisure and recreation that would appeal to a mass market.

How much more sensible to promote the health-giving properties of seawater over those of mineral springs: so much more of it and the sands too. Of course, a certain business acumen was necessary and full-scale development of the coastal resorts had to wait on the railway age. Scarborough (photo 56) makes an interesting link between the two eras. The first seaside resort, it began to grow around 1700 when mineral springs at the base of its sea cliffs were popularised. Once the benefits of the sea came into fashion Scarborough made an easy transition and became one of the most popular seaside towns.

Opportunities for urban recreation were also developed in the nineteenth century. Under Queen Victoria crowded and insanitary conditions in the towns and cities translated through the values of the time into a drive, led by Prince Albert, for higher standards of public health and opportunities for education. The urban scene was transformed by better housing, sewers, schools and hospitals and intertwined in the 'new order' there was a deeply held view that there must be space and provision for self-improving leisure and recreation. In many towns there was expression of civic pride in a proliferation of formal parks, museums and galleries

56 Part of the South Sands, Scarborough, North Yorkshire. The promenade (Golden Mile) which runs along them is at the heart of the town's traditional seaside entertainments and amusements. There are 154ha of parks and gardens especially on the steep slopes above the shore. The original Spa is located just above the sands in the middle of the photograph and to its right is the Ballroom Theatre and Grand Hall. Today Scarborough attracts nearly one million residential visitors each year.

that remain popular today. Cathays Park (photo 57) and the Civic Centre in Cardiff provide a late and excellent example, but wherever the like exists the model is generally Hyde Park and the huge museums complex in Kensington. The purpose was to provide a somewhat contemplative and sedate setting to experience fresh air and nature and to impress on the minds of ordinary people, for their benefit, the great achievements of the age. In other places the recreational role was combined with more serious ends as at Kew or the smaller 'Botanic' gardens like those in Cambridge, Sheffield and Edinburgh.

Countryside recreation also started in the nineteenth century but it was slow to catch on and again depended on a rather elitist appeal. Wordsworth, though he sought to keep working-class hoards away, unwittingly popularised the Lake District and when the railway came to Windermere the crowds came too. What would he think of Grasmere (photo 58) now, where Dove Cottage, his former home, is one of the principal attractions? In another sense, if not in the scale of recreation use, the nineteenth century was the vital period that cast attitudes and an approach to countryside management that binds us still in attempts to

57 Cardiff, Cathays Park, South Glamorgan. The 24ha setting of Wales' Civic Centre is surrounded by imposing public buildings in Portland Stone. In the bottom right-hand corner of the photograph the complex is dominated by the City Hall (1901–6). On the left are the Law Courts and to the right the National Museum of Wales, opened in 1927 by King George V and intended 'to teach the World about Wales, and the Welsh people about their own fatherland'. Behind the Museum are some buildings of the University of Wales, and the double circular colonnade is the Welsh National Memorial and Cenotaph.

resolve modern conflicts. The eighteenth and nineteenth centuries saw massive change in the countryside (depopulation, enclosure, and new agricultural practices) and by the late nineteenth century there were the first murmurings at what was being lost. But from the beginning concern came from two sides. On the one hand there were those interested in the aesthetic appeal and amenity of the countryside and eager to protect and promote access for the pleasure of ordinary people. On this side the romantic movement and the support of leading figures such as John Ruskin, William Morris and Octavia Hill were important. Their influence was crucial in the emergence of supporting organisations, the Commons, Open Spaces and Footpaths Preservation Society (1865), Britain's first rural amenity society, and the National Trust (1895), for example. On the other there was an incipient conservation movement aiming to protect the natural environment in the cause of science and which called for nature reserves free from public access. The discoveries and contemporary prominence of Darwin and Huxley gave much impetus and the conservationist viewpoint was also promoted through various societies, the Royal Society for the Protection of Birds (1889) and the British Ecological Society (1913), for instance. The early separation of conservation and recreation lobbies has done much to foster conflict and inhibit an integrated approach to countryside planning and management ever since.

58 *opposite* Grasmere, Cumbria. Although no more than a village (population 850), location in the heart of the Lake District and historical associations with Wordsworth ensure that Grasmere is a thriving tourist centre. It is a convenient stopping off point for coach tours as well as other visitors and has no less than nine AA listed hotels (six rated three star and above) and is correspondingly well provided with guest houses and bed and breakfast places. In 1987 the National Park information centre in the town dealt with over 92,000 visitors drawn into the honeypot, especially on a wet day, by numerous cafés, gift and craft shops, specialist climbing outfitters and Dove Cottage itself (80,400 visitors in 1987).

COUNTRYSIDE RECREATION DEMAND

When recreation demand first began to increase rapidly the scale of the new phenomenon was largely unexpected and not understood. There had been hardly any research. But by the 1970s and thereafter leisure studies abounded and any shortage of facts and figures about countryside recreation was more than rectified.[6] Although over the years some developing trends have intensified and others abated the basic patterns of countryside recreation as they affect different kinds of people and are variously distributed over space and time were soon described and, except in scale, have changed little in the meantime.[7] In contrast over the next 20 years changes in the pattern of demand may be substantial.

On a typical summer Sunday around 18 million people visit the countryside. For those wanting open spaces the National Parks are particularly popular, the Peak District alone accounts for 20 million visits per year. Others may plan a day out around a more specific objective and the National Trust leads the trend. Between 1955–87 visits to the Trust's properties increased by 6–7 per cent compound per annum and now stand at 8.5 million. These are just two examples from an extremely diverse spectrum (see diagram below). Participation rates are high and the countryside appeals to just about every type of person. Participation is similar for men and women and their interest is sustained throughout life. It also draws people from all social groups, only the least well off are under-represented and even amongst these – the semi-skilled and unskilled manual workers – 35 per cent manage a countryside visit in a typical month.

Looking ahead to the year 2000, the Countryside Commission believes that recreation demand through the public's enjoyment of an attractive countryside will grow in importance.[8] However, the period of explosive growth in demand is almost certainly over. Countryside use has increased with affluence, but the

Countryside recreation activities. (Source: based on Countryside Commission, *Recreation 2000*.)

Unmanaged countryside (71%)

Managed countryside (29%)

		%
1	Drives, outings, picnics	19
2	Long walks	18
3	Visiting friends, relatives	14
4	Sea coast	8
5	Informal sport	12
6	Organised sport	7
7	'Pick-your-own'	4
8	Historic buildings	4
9	Country parks	4
10	Watched sport	3
11	Other	7

59 *opposite* Barton Broad, Norfolk. One of the largest and most popular Broads it was long famous for its clear water and rich aquatic flora and fauna, but it is located in the upper reaches of the narrow River Ant and as boating holidays became more popular in the 1960s and 1970s congestion became a serious problem and environmental change set in. Today's holidaymakers find a much impoverished environment, but although the congestion remains in the peak season (as shown by the large number of boats at Barton Staithe) boating is not the main cause of environmental deterioration. The damage has been linked to sewage discharges and phosphate pollution and restorative work is now bringing the Broad back to life.

closest link has been with rising car ownership and increased availability of free time. Three-quarters of countryside trips are made by car (a mere 5 per cent by public transport) and recreation demand has risen alongside car ownership. There were only 2.25 million cars in 1950, but 5.5 million in 1960 (30 per cent of households) 12 million in 1970 and 15 million in 1980 (58 per cent of households). Today there are 17 million (60 per cent of households). In similar vein greater flexibility soon came in after the war as the six-day working week rapidly gave way to a five-day week and people gained extra holiday entitlements. In 1951 only 3 per cent of male manual workers got more than two weeks' holiday. By 1981 more than four weeks was commonplace (35 per cent) and many were entitled to more than 3 weeks (47 per cent). Taking these two trends together the rate of increase in recreation demand might be expected to ease off, and that is in fact the case. For example, visits to the North York Moors National Park numbered 11 million by the mid-1970s, but are still about the same today. Nonetheless, demand looks set to continue to be the single most important consideration in

recreation management and planning, for the next two decades at least. The 'boom' may be over, but there remains some scope for additional pressure from growing numbers of countryside users and more importantly there are developing trends pointing to a broadening of pressures on the countryside through changes in the pattern of impacts.

Somewhat surprisingly the countryside has so far been spared the worst consequences that could reasonably have been expected to follow on so much recreational use. Hitherto, people have consistently chosen to arrange their recreation in ways which lead to a very marked concentration of activities in space and time. Even though the car is so important most visits involve surprisingly short distances, less than 25km in a third of cases. The Peak District is the nearest National Park to south-east England, but its extraordinary popularity is due to its accessibility to adjacent areas, Manchester, Sheffield, the Potteries and West Yorkshire, rather than populations further afield. People tend to return time and again to a few familiar places rather than seek out new haunts, and in doing so they are inclined to keep to the main routes. At journey's end they tend to congregate with others at a smallish number of main sites and attractions, the so-called 'honeypots' and rarely stray more than a few metres from their cars (see photos 58, 61, 62 and 63 for example). In timing, too, distinct peaks occur. On a seasonal basis July and August are the most active months (even May, June and September are much quieter). Recreation is also concentrated on particular days. Sundays may attract up to five times as many visitors as ordinary weekdays and numbers tend to reach their maximum in late afternoon then decline suddenly when people decide to go home, usually around 5 o'clock. Time after time studies of different sites have confirmed these patterns. In the Norfolk Broads (photo 59) for instance, several surveys have recorded over 1,000 boat movements per day at Horning on the Bure on Sundays in August. On the Yare and Waveney traffic is reduced by 50–80 per cent at the same time and in all cases the weekday traffic flows at only about 75 per cent of the Sunday peak.

60 *opposite* Stonehenge, Wiltshire. This great stone circle surrounded by a ditch and earthwork 90m in diameter is Britain's most famous prehistoric megalithic monument. It is also the most visited (around 500,000 per year) the pressures being exacerbated because (to the surprise of many seeing it for the first time) it is far from remote in the middle of Salisbury Plain – the circle is directly on the busy A844 near to its junction with the main A303 trunk road. Sadly the popularity of the monument is its downfall and for protection it is now fenced off and approached somewhat incongruously via a subway (the entrance is clearly visible in the photograph) connecting it to the large car park.

61 Snowdon, Gwynedd. As the highest mountain in Wales, Snowdon attracts upwards of 250,000 visitors per year. The rack and pinion railway from Llanberis (opened in 1896) carries up over 1,000 people per day in the peak season, but on the busiest days almost twice as many walk. The summit and the approach paths are heavily eroded and in 1979 a management programme sponsored by the Countryside Commission and National Park Authority was set under way. In places the restoration work is unsightly – gabion cages on parts of the Pyg Track for instance – but the first priority is to protect the mountain. One important effect stemming from the popularity of Snowdon is that it serves to ease pressures on surrounding – equally fine though lower – peaks and though a railway and summit café seems inappropriate to some (what chance of planning permission today?) it brings the openness of mountain heights to thousands who would otherwise never see them.

62 *opposite* Twyni Bach (Ynyslas, sand dunes). The private car has made the countryside accessible and most visitors remain content to follow the throng, carrying along as many of the conveniences of urban living as possible, and at the end point of their journey tend to prefer not to stay more than a few yards from their cars (if they get out at all). Here people congregate near the access point onto the beach even though they could separate themselves much more widely and are mainly picnicking near their cars or in the dunes, but the dunes are easily eroded as people clamber about and children slide down their mobile slopes.

Since total demand in terms of numbers of people has levelled off, the really interesting questions over the next 20 years will be whether and how the pattern and impact of use so firmly established in the past might change. Will users' surprisingly conservative behaviour and its impacts remain so confined in future? Almost certainly not. Changes in age structure (the population is ageing quite rapidly) and earlier retirement will probably lead to new recreational demands. Further expansion of the middle class, associated with a changing employment structure developing alongside the new technologies, for example, will also have an effect. Against a background of affluence countryside users represent an irresistible market and further commodification of recreation activities and a push given to demand is inevitable. People who already use the countryside will use it more often and as they gain in knowledge and self-assurance (consider the effect of the vast range of guides and other books that are now sold) are likely to be more adventurous and discriminating; searching for new and more satisfying experiences their impact will be felt in new activities over a wider area and at new times. The Countryside Commission has found that frequent users know most about the resources available for countryside recreation, but as yet they are a minority (around 25 per cent of all users).[9] As this group grows in importance the Countryside Commission will need to face more overtly the smouldering paradox inherent in its own role and remit. The Commission is statutorily obliged to develop policies which both conserve the countryside and promote its enjoyment by the public. On the one hand, the Commission cultivates an air of preparedness and responsiveness in the face of the rising tide of pressures. On the other, some of its recreation policies belie an invitation to accelerate these very same pressures which must, of course, bring implications concerning conservation and the natural beauty of the countryside. Footpath policies and some of the Commission's own figures illustrate the point well.[10] The footpath network stretches to a huge 190,000km but only 34 per cent of countryside walking is on these rights of way (33 per cent is on roads!) Few people own maps (even of the simpler kind?) and only 15 per cent are able to or normally use them in their recreation. Yet somewhat contrarily the Commission regards the footpaths as the 'single most important means by which the general public can enjoy the countryside'. Does the Commission mean that all those people who want only to set up a deck chair and picnic at the side of their car (photo 62) ought to be pounding along the footpaths instead? In any event the Commission intends to promote the paths. Clearly, there is enormous scope to disperse visitors from the car parks and main attractions, and there may be great benefit from enhancement of opportunities for personal enjoyment of the countryside, but impacts consequent upon the new pattern of use may be substantial. Time will tell and show to what extent positive or negative policies are used to intervene.

PROVISION FOR COUNTRYSIDE RECREATION

The broad public interest in recreation and conservation is recognised in protection of the countryside and provision of recreational opportunities through statutory intervention. Governments have been particularly responsive in the face of a coincidence of post-war pressures involving development, agriculture,

forestry and recreation and the rise of environmentalism (notice the explosive growth in membership of amenity societies), but the roots of concern reach much further back, to the Victorian era at least. There was a major reassertion of Victorian concerns in the late 1930s apparent in growing unease about new problems of afforestation and urban sprawl, particularly ribbon development around major towns and cities, and legislation seemed imminent. Unfortunately, recession and war intervened and the main framework of laws and regulations affecting land use controls in the countryside was necessarily delayed. A flurry of legislation followed immediately after the war and in the intervening 40 years statutory provisions have been steadily extended and updated notably with two specific acts, The Countryside Act 1968 and The Wildlife and Countryside Act 1981, dealing with countryside matters.

At the centre of the provisions introduced in the 1940s were the twin concepts of National Parks and National Nature Reserves (NNRs). The path leading to the legislation in its final form is significant. Two important reports, the Dower Report 1945 and Hobhouse Report 1947,[11] recommended establishment of National Parks having regard to amenity and natural beauty and opportunities for recreation over extensive areas of unspoilt countryside and to achieve this end envisaged planning on a relatively broad scale. Recommendation for 12 areas with National Park status followed and in time, if we include recent special provisions affecting the Norfolk Broads, all but the South Downs were designated. A report by the Wild Life and Conservation Special Committee (Huxley Report) on ecological and scientific aspects of nature conservation was also published in 1947 and proposed over 70 NNRs taking in around 30,000ha and powers to acquire important sites for their protection and scientific study.

Working in a considerable spirit of cooperation these Committees sought complementary powers and similar operational structures, both groups proposing central (i.e., national) executive bodies, respectively a National Parks Commission (NPC) and Biological Service. The point was to come forward with relatively well integrated proposals that would meet their main aims and at the same time reflect a balance of interest where their principal objectives differed. Most important in this respect was the concept of Conservation Areas and a joint proposal that there should be more than 50 such areas covering about 16,400sq km. In these Conservation Areas the aim would be to maintain landscape beauty, scientific interest and recreational opportunities. Therefore, in the Conservation Area concept there was a crucial recognition of interdependence between the interests of amenity and recreational use of countryside and nature conservation.

However, again in 1947, the Town and Country Planning Act 1947 placed planning control in the hands of Local Authorities so creating a precedent that was necessarily followed in the National Parks and Access to the Countryside Act 1949. In consequence although a National Parks Commission was set up it was not executive; it could designate National Parks but thereafter only advise Local Authorities and could not own or directly manage land nor carry out research. Nature conservation on the other hand was pushed forward as a scientific rather than a planning matter and treated somewhat differently. In 1949 a Nature Conservancy was established as a national agency with executive and advisory powers. It could own and manage land and carry out research, but the idea of Conservation Areas was dropped. Instead, provision was made for Areas of Outstanding Natural Beauty (AONBs) to be designated under the powers of the NPC. The upshot is that in both a structural administrative and conceptual sense the separation of amenity and recreation and nature conservation that originated in the nineteenth century became institutionalised and the recognition of their interdependence that was so desirable was lost.

Under the new legislation designation proceeded swiftly and was by far the greatest achievement in the period through to the late 1960s. Certainly in the case of the National Parks there was, in the lull after the war, an unrepeatable opportunity to provide a formal basis from which to defend the countryside; had it been missed it seems extremely doubtful whether circumstances would ever again have favoured designation. The first National Park, in the Peak District,

63 Oxwich, West Glamorgan. The Gower Peninsula, although very small, has remarkably varied and beautiful habitats and landscape – 22km (40 per cent) of its coastline is of national significance and mainly held by the National Trust. Gower has been protected from development by historical inaccessibility and, more recently, by AONB status, but visitor pressure is now enormous. The dense populations of the South Wales coalfield are nearby and Swansea (pop. 160,000) is literally on the doorstep. The full spectrum of problems and conflicts inherent in trying to reconcile recreation and nature conservation come together at Oxwich, but are overcome by close management. Long popular for its wide sandy beach, traditional pursuits are now supplemented by sailing, power boats, water skiing and wind surfing, the recreational demands coexist alongside a splendid National Nature Reserve extending to 240ha under control of the Nature Conservancy Council. Access is carefully regulated but visitors are accommodated and their enjoyment enhanced by a well maintained network of guided walks, excellent interpretation boards and the facilities of a reserve Information Centre.

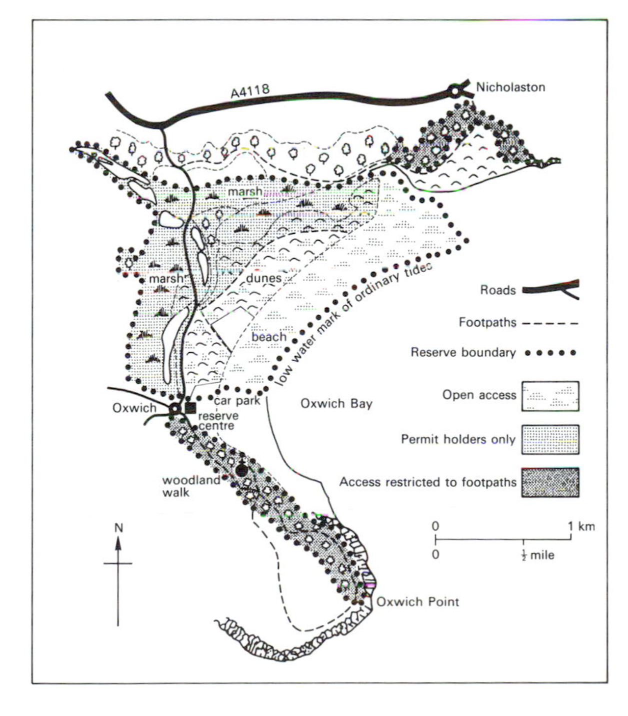

Oxwich, West Glamorgan. (Source: based on Nature Conservancy Council, *Oxwich National Nature Reserve*, 1979, 1981.)

was designated in 1951 and the remaining nine followed quickly by 1957. In 1956 the Gower Peninsula became the first AONB (photo 63); there were 21 by 1966 and there are 38 today. A little later, but with direct provision for recreation in mind, the NPC moved to designate long-distance footpaths though the first, the Pennine Way (photo 64) was not completed until 1965 and others that were planned, like the Cambrian Way through mid-Wales, never overcame opposition to their formation. There are now 13 routes totalling 2,699km and the Thames Path and extensions to the South Downs Way are nearing completion. The Nature Conservancy established over 80 NNRs by 1966, 224 today, and over 1,400 SSSIs, 4,729 today, though in the case of SSSIs there is the disappointment that a high proportion (possibly 75 per cent) have been seriously damaged, and in many cases ruined since being declared.

64 Pennine Way, Greater Manchester. Contrast in extremes. The scene is dominated by the geometrical swaithe of the M62 graphically illustrating the way that modern communications may reshape the landscape. Here, just south of Blackstone Edge, the Pennine Way is routed over the motorway via a concrete bridge (middle of photograph).

In other respects optimism that may have been fuelled by the creation of the NPC and Nature Conservancy was not borne out. As demand for recreation exploded in the 1960s people became increasingly aware that an important public role in provision and management was not being met and that the planning system was being expected to take on problems with which it was ill-equipped to deal. Local authorities did not have the powers, professional expertise or resources to respond and the lack of a national coordinating body meant that there was insufficient consistency in recreation and leisure policies between local authorities up and down the country. The Countryside Act 1968 was intended to be the remedy. The overall thrust of this Act was to strengthen the position of conservation in public policy but interestingly it was conservation in the planning tradition of amenity and natural beauty that was stressed rather than nature

65 Tissington Trail, Derbyshire. The trail follows the disused Buxton to Ashbourne railway line for over 20km from Ashbourne to Parsley Hay providing a graded footpath and cycleway through some of the most attractive countryside in the Peak National Park. There are way marks, picnic sites, ranger services and cycle hire facilities.

conservation.[12] The NPC was replaced by the Countryside Commission, a new body with a *national* advisory remit and powers closely modelled on the NPC but with the addition that it was able to grant aid purchase of land for recreation by local authorities anywhere in England and Wales and could carry out research and experiments. The effect of the act was to enhance the role of recreation management and draw it closer to the objectives of nature conservation where they seemed to be in parallel. Perhaps the most significant impact of the new powers and resources in this respect came in the concept of Country Parks, areas of say 80–120ha on the urban fringe that would be accessible to large numbers of people, able to stand heavy use and deflect pressures away from the National Parks or other popular and/or sensitive resources. Active management of a more specific nature was also widely introduced (photo 65). The Commission has used its powers of research and experiment to introduce and monitor a whole host of 'projects' on, for example, footpaths, signs, way-marking, wall-rebuilding and tree-planting.

CONFLICT AND MANAGEMENT

Recreation in the countryside is generally resource-based; recreational opportunities are limited by the availability of the resource itself, and although recreation pressures and demands now necessitate some protection and positive management, the special character of the countryside is quite independent of recreational use. It is valued for its own sake and recreation is usually incidental since landowners tend to see the main purposes of countryside in other ways. On the broader scale recreation has typically been accepted alongside other major types of land use and it would be splendid if countryside could always be useful *and* open to public access, but of course this is not so. From time to time conflicts have inevitably occurred and the extension over the countryside of planning and other legal and administrative controls, coinciding with growth in recreation demand, has fulfilled a major role. Many would argue that conflicts, if not wholly resolved over the last 40 years, have at least been contained, but any success achieved has depended to a considerable extent on the particular problems around which conflicts have developed. Major public conflicts first became apparent over matters of principle: the question of public access for recreation that first flared up with a mass trespass on Kinder Scout in the late 1930s is a good example. Recent conflicts have involved more pragmatic issues. It is convenient to divide these conflicts into three types, though in reality the neat distinctions are somewhat blurred. In each case the same broad interest groups seem to emerge in adversity and the issues that are crucial as the process is worked towards a conclusion tend to recur and overlap. The first type of conflict revolves around the perennial difficulty that some farmers always resist the extension of access for recreation only to find themselves challenged by a well-organised access lobby. That is an issue in itself, but the result, together with the fact that the availability of land suitable for recreation is limited, is a concentration of recreation pressures in certain areas. These pressures result in conflicts with conservation and amenity interests and between different types of recreational users. A third set of conflicts involve specific major development proposals, but again it is questions affecting agriculture, conservation, amenity and recreational use that come to the fore.

Access

In terms of recreation the access question is fundamental. Only about 13 per cent of Britain is publicly owned and much of this land is held by bodies such as Water Authorities, the CEGB, British Rail, British Coal, and the Ministry of Defence, so even on 'public' land access cannot be taken for granted. It is the remainder that is privately owned that makes up most of Britain's diverse countryside and the main recreation resource. Unfortunately, the relationship between landowner and visitor is uneasy, even hostile. Members of the access lobby, the Ramblers' Association, for instance, bridle against what they see as the excessive power of landowners to impose unreasonable restrictions on people's freedom to use the countryside simply because they own it, especially when owners employ tactics calculated to deter in places where there are public rights of way. In a recent survey the Countryside Commission[13] has highlighted both sides of the problem. The Commission found that many rights of way are obsolete, neglected or obstructed (mainly by being fenced or ploughed out) and that dubious 'Keep Out' signs and other similarly anti-social actions are common. Amongst farmers anti-visitor attitudes were the norm. Only 6 per cent thought there should be more public access in the countryside and less than a third went along with the idea that farmers have a general responsibility to society to allow access to the countryside. From the farmer's point of view the public cause problems. Some 71 per cent complained that their land is used without permission (trespass) and cited damage (to livestock, crops and hedges, for example) as a significant problem. There is vandalism, gates are left open, and litter is left behind. These problems and the attitudes they foster are important.

The basis of the conflict and differing interests of the protagonists are more or less clear-cut, but it is interesting to consider how their rival perspectives are brought to bear on the problem and the way in which it is managed. The conflict manifests itself on two levels, through formulation of policy on the general issue of access on the one hand and, within the framework so determined, on the more practical level of day-to-day recreation management. The most significant recent legislation in this area has been the Wildlife and Countryside Act 1981 and Adams has provided a full and accessible account chronicling bitter argument and the protracted stages through which the bill passed before becoming law.[14] The crux of the matter here is that the conflict operates through highly political processes resting on the relative power and resources of different interest groups and their skill in influencing the shape of the policies that emerge. In this conflict pressures from the access lobby must be taken seriously, but it is the interests of property and landownership that are dominant. Shoard collates some interesting evidence.[15] Members of National Park Authorities, for example, must be appointed two-thirds locally and one-third by the Secretary of State. Not surprisingly then most tend to live locally and rurally, but it is the representation of farmers that is revealing. One study showed that in 1985 farmers and landowners had cornered a third of the places on the three Welsh National Park committees. But the problem goes even deeper for a study of 1983–4 showed that no less than 40 per cent of National Park members appointed by the Secretary of State were farmers or landowners. Other interests such as tourism or conservation are poorly represented by comparison and when amenity interests do find a place in the membership it is the 'establishment' organisations like the National Trust that are successful, campaigning groups such as Friends of the Earth and the Ramblers' Association are not welcome. The same sort of pattern is found on the Forestry Commission, Nature Conservancy and the Countryside Commission.

On a more practical level recreation conflicts are worked out in several ways. Where there are new recreation demands or specific development proposals there may well be concern, understandable enough, amongst affected farmers and other local people. Where they are able, farmers will often try to avoid the problem by simply opposing recreational use. In Wales, for example, where the influence of farmers is considerable, the Countryside Commission spent a decade trying to win acceptance for a long distance footpath to run up through mid-Wales, but eventually succumbed to local opposition; in 1982 all plans for a Cambrian Way, as it would have been, were dropped. In another instance, with a rather different twist, the Snowdonia National Park Authority found itself caught up in unwelcome publicity caused by differences between farmers and walkers around Llanuwchllyn and Dinas Mawddwy. The farmers became dissatisfied with the numbers and behaviour of walkers making *free* use of the Aran mountains and grouped together to close off their land and deny access to the tops (which they were perfectly entitled to do). When recreation is resisted in this way little is gained. The pressures are simply displaced elsewhere and the whole conflict is quite likely to be re-enacted. In any event such attitudes make a poor foundation on which to build future partnerships. The Wolds Way is a sad case in point. It should have been a triumph for the Countryside Commission, but alas no. The Way was designated in 1982 and unlike most other long distance footpaths passes through good farmland, and therein lies the rub. The route, 127km from Filey to the Humber, was hard fought over by farmer and walker alike. Neither group is particularly satisfied with the final outcome, and in terms of human relationships designation was bought at a high cost that will sour the achievement for years to come.

Although these few examples are not typical it is difficult to generalise about the reception that new proposals will meet. Designation of the North Pennines AONB was strongly contested. There was concern about damage to the local economy and there was sufficient momentum in the local opposition to force a public inquiry – the first affecting AONB designation. It was not until 1986, eight years after the Countryside Commission brought the proposal forward, that the Government finally confirmed AONB status for 85 per cent of the recommended area. It became Britain's 37th and largest AONB. At the other extreme the Howardian Hills AONB, about 130 square km of North Yorkshire around Castle Howard, was confirmed in 1987 and became the 38th AONB just six months after the Countryside Commission announced the intended designation.

Carrying capacity

For many visitors the essence of countryside recreation is to move into wide open spaces and away from crowds of other people, and on first consideration this aspiration seems to fit in well with the resources available. The importance of the countryside is recognised and there is the responsibility to preserve and enhance its natural beauty and make it available for public enjoyment. There is no doubt that informal countryside recreation is enormously popular. But it is the scenery that is available. Relatively little of the countryside is open for people physically to use, the more so because although the resource seems extensive its capacity to absorb recreational activities is quite limited (photo 66). Capacity is soon reached even at apparently modest levels of use and conflicts are increasingly common. The problem is that for various reasons, the sensitivity of the countryside itself, for example, and because there are access problems that tend to be exacerbated because people are unadventurous and lacking in the necessary knowledge and confidence to make the best use of what is available, excess pressure develops at the better known and most popular places. The conflicts have social as well as physical dimensions, though the latter tend to receive a rather more overt attention in recreation management where the natural environment is threatened, say, by some particular level or type of use. The Three Peaks area of Yorkshire presently suffering erosion from over use by walkers is a case in point. The damage is most severe on a 40km section used for the challenge walk over Whernside, Ingleborough and Pen-y-Ghent. The full route is followed by at least 15,000 people per year. Ingleborough, approached over vulnerable peaty soils is the main peak and bears the brunt of the pressure. It is climbed by about 120,000 people every year and in places the 'path' is 30 to 40 metres wide. Other examples are shown in photos 61, 62 and 63.

The nature of social conflict is much more subjective. Perhaps the most intuitively familiar and easily recognised problems are congestion-related. 'Honeypot' locations, for example, come under particular pressure and many a day out does not live up to expectations or is spoilt by lack of foresight and poor planning. Often when people pick a destination they simply do not anticipate the extent to which hoards of others may have the same idea and upon arrival it is found that the chosen setting cannot be used and enjoyed in quite the way that was hoped and intended. Naturally, such experiences are disappointing. In the Lake District, a particularly beautiful part of Britain, visitor pressure is enormous and the most popular centres like Grasmere (photo 58) suffer severe congestion. Grasmere has a magnificent bowl shaped setting right in the heart of the area. It is mid-way between Windermere and Keswick on the main route through the National Park and Ambleside is nearby. It is an excellent base for fell walkers with Scafell Pike (England's highest mountain), Helvellyn and others on hand. However, Grasmere's particular claim to fame which reinforces the natural advantages of setting and accessibility is that William Wordsworth chose to live there. With origins in the nineteenth century a whole tourist industry has been developed around the image of his writing and life in the village.

Plainly, there are problems with this sort of congestion. Grasmere makes a good example but the problem is reproduced in other places and not at all uncommon, so it is interesting to consider the question of management. The distinctive thing about people who head for places like Grasmere is that they share the same general purpose and want to do the same sorts of things. There is no particular

social conflict between them, just the problem that on some particular day too many people set off to do the same thing in the same place causing some to feel *crowded* and disappointed.[16] But the problem is transient, the disappointment passes and if the visitor comes back when conditions are different there may be no experience of crowding. In management the first point to notice is that application of the honeypot concept has been seen as an important device to stay pressures on more sensitive places. Second, even if the level of use in some place where the policy operates does start to cause concern, it may be extremely difficult to ration access to the honeypot as positive powers to achieve that end are not usually available – which is why the recreational use of such places as Grasmere grows with so little constraint. The managerial response may well be to regard the problem as a superficial one and treat it as self-limiting. If feeling crowded is a transient experience people who prefer visits at less busy times will learn what to do next time. Perhaps it is not unreasonable to expect them to accept displacement into off-peak times or seasons, even other places, or to notice that good weather brings out the masses?

Other problems are more serious and do involve social conflicts. People who use the countryside during their leisure decidedly do not always share the same general purpose or see the importance of countryside and its recreational utility in the same way. These deeper social conflicts occur because people obtain their satisfaction from the countryside in different ways, often have different values, and as a result may regard each other's recreation activities and behaviour as mutually exclusive. There are some well-known examples. In the Lake District there have been problems centred on the multipurpose use of Lake Windermere for water-skiing, different types of boating and more passive waterside pursuits. Along the Ridgeway, about a 150km Long Distance Path in southern England, walkers and horseriders are involved in a major conflict with cross-country motorcyclists. In the Norfolk Broads (photo 59) there have been complaints for more than 30 years about angling and the holiday hiring of motor cruisers. In fact motorised versus non-motorised recreational use is a common theme that tends to divert management from the more important underlying social basis of many of these conflicts; if there is a managerial response it tends to be concocted on an *ad hoc* basis and the usual course is to appeal to people to use commonsense and respect a sense of fairplay and, where this fails, to introduce zoning (separation of activities in space and time) rather than look for some deeper solution based on understanding the nature of the conflicts involved. Zoning has been favoured for Windermere, and on the Ridgeway the Countryside Commission has been seeking exclusion of motorised traffic along a 60km byway stretch of the route on Sundays and Bank Holidays (1 May to 31 October only) since August 1986, but without success. The Ridgeway is an ancient road and in any case the Trail Riders Fellowship shift blame onto agricultural vehicles. They also feel that criticism of their pastime, 2 per cent of Ridgeway use they claim, is exaggerated. No doubt walkers and horseriders wonder why 2 per cent should be allowed to spoil the fun of 98 per cent. In the meantime, a voluntary code of conduct is being distributed amongst motor vehicle users.

Such policies are inclined to seem plausible and tend to win strong support (except amongst any who are directly affected and lose out) because they are

66 *opposite* Derwent Reservoirs, Derbyshire, looking north. To create these three reservoirs, Ladybower (foreground), Derwent (middle) and Howden (background) the village of Ashopton was evacuated and flooded over. They were used as a practice zone before the World War II Dambusters' raid. Nowadays they are an important recreational as well as water supply facility. Water draws people for informal recreation and the 'gritstone edges', Bleaklow, Howden and Derwent, provide some of the finest walking in the Peak National Park. There is a public road to the northern end of Howden reservoir and on both sides of Derwent reservoir below Howden Dam, but a visitor management scheme operates over weekends and Bank Holidays between Easter and the end of October. At these times motor traffic has access only as far as Fairholme car park (just below Howden Dam) and then only along the western side of Derwent reservoir.

highly amendable to quasi-pluralist rationalisation through the taken-for-granted idea that 'we must all live together' in an atmosphere of compromise and consensus. Unfortunately, this approach ignores the deeper basis of some social and psychological conflicts and through that omission may well exemplify, albeit unconsciously in most cases, a model of non-decision taking in which responsibility for positive resource management is avoided with consequences for both recreation users and the resource. In the Norfolk Broads, for example, the basis of the conflict between anglers and boat users was apparently simple; there was obvious degradation of the environment together with an expressed opinion from anglers that the quality of fishing was declining, and both coincided with a considerable post-war increase in the popularity of the area, particularly for holidays by hire motor cruiser. The managerial response to the problem followed a familiar pattern. At first there were meetings with users' representatives and calls for more understanding and tolerance from all concerned, then there followed a very limited scheme voluntarily to restrict boat movements over about 5 per cent of the waterway before 9.00 am on Sundays (June to October only). But subsequent research on the recreational use of the Broads[17] has shown that the conflicts are far more complicated and that actions stressing better behaviour and zoning may not achieve the improvement in user satisfaction that is intended or make best use of the area's resources. The measures introduced in the Broads were a purely pragmatic reaction by managers to a well-orchestrated campaign led by a local anglers' association; an example of the political behaviour with which managers become bound up. Careful investigation in the Norfolk Broads showed that there is an interesting pattern of reciprocated and unreciprocated conflicts (see diagram below) some of which have a strong social basis while others are weaker or limited

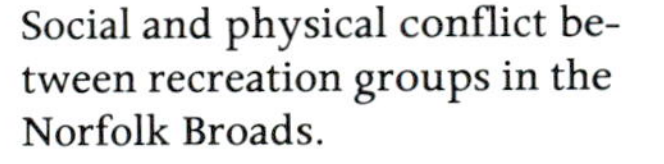
Social and physical conflict between recreation groups in the Norfolk Broads.

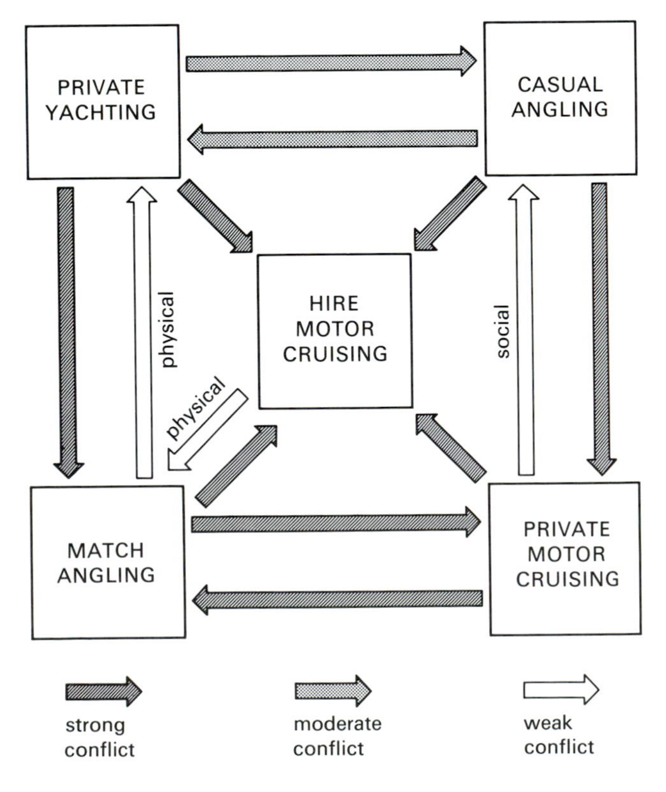

to physical effects. Unlike crowding the feelings associated with social conflict were not transient, they developed gradually and cumulatively from a series of unsatisfactory outings and then persisted from one visit to the next. From a cursory look at the Broads it might have appeared that conflict was decreasing. Many visitors had started using the area comparatively recently and for many years the trend had been for the number of visitors to increase. However, closer investigation revealed that the strongest conflict was felt by the most experienced and/or committed users, particularly amongst anglers. Many of these people had in effect been displaced; they reported visiting the area less than they used to and were often interviewed at off-peak times and places rather than the more popular locations they said they preferred. The consequences implied for both user and resource are important. The aim in positive management is to preserve and enhance the countryside and at the same time be able deliberately to allocate resources to their highest potential for providing recreational opportunities and satisfaction. The pragmatic approach exemplified in the Broads is no more than an after-the-event reaction to effects of conflict rather than an attempt to understand and anticipate it. If conflicts are viewed as self-limiting, say through replacement of existing users by more tolerant newcomers so that the apparent conflict subsides, then recreation managers ignore the likelihood that over a period of time both the resources and human experiences will suffer decline. Should users whose recreation does not make very specific resource demands be able to displace others whose activities are highly resource dependent? What is the effect on the quality of recreational opportunities and satisfaction obtainable from the resource? How is the natural environment affected by the new level and type of use? Questions like these should be positively addressed in recreation management whereas non-decision-taking encourages the lowest common denominator and a more uniform standard of provision in terms of both the quality of the resource and recreational opportunities.

Economic development

Pressure for development produces conflict of a different nature. There is a long history of opposition to certain specific major developments in the countryside. The moves against the Trawsfynydd nuclear power station (photo 67) in Snowdonia during the 1960s stand as a good early example and gave a foretaste of larger and more sophisticated environmental campaigns that would follow. In retrospect it is interesting that at the time objections against the technology were overshadowed by the issue of whether or not that type of development could be appropriate *in a National Park*. There is room only to pick out and discuss a few contemporary examples, to give an idea of the type of controversy that exists.

The Town and Country Planning Acts provide the fulcrum around which the various conflicts have been played out and have been seen by environmentalists as the principal force through which to defend the countryside. But it *is* a defensive stance, planning operates through negative powers offering the possibility to resist development which seems undesirable, but unable effectively to promote change in a positive way. What is more, there has been little direct control of agricultural development. Agricultural buildings for example are largely

outside development control powers, so the role of planning has mainly been to influence the shape and extent of other developments and in particular to keep out the worst impacts of urban development that might otherwise have crept further into the countryside.

The reasons for conflict over road building are easily appreciated – is there 'need' for the road, what is the 'best' route, is the cheapest solution the 'right' solution? – but are quite intractable and recur time and again. Major roads sever the landscape irreversibly, perhaps none more so than the M62 which was cut across the Pennines (photo 64) in the 1970s after a long campaign. Environmentalists can point to some victories, for example the route of the M40 has been substantially altered to avoid important wetland habitats, but rather more defeats. In this last category the decision on the Okehampton bypass was particularly shameful. After more than 20 years of complaint about congestion in the town a Joint Parliamentary Committee accepted the case for a bypass, but recommended that a route must be found which avoided Dartmoor. Instead the Secretary of State overruled this view and the road is being built to cut through the National Park to the south of the town. Controversy currently surrounds the M3 Winchester bypass. The Countryside Commission opposes the Department of Transport's preferred route on the grounds that it cuts the chalk downs east of the city, so severing an AONB, and would destroy part of an SSSI and several ancient monuments. The Commission believes it can suggest a route that is no more expensive and which avoids these environmental problems.

Conflicts with quarrying stem from the geological coincidence that rocks that are economically important often happen also to be the basis for much of Britain's most beautiful scenery (see diagram below). Limestone is one of the best examples. Limestone quarries scar the landscape, make noise and dust and push heavy lorries

67 *opposite* Trawsfynydd, Gwynedd, showing the nuclear power station as it was when opened in 1965. Construction required 3,800 cubic metres of excavation leaving an extremely incongruous and dominant intrusion on the landscape of Snowdonia. However, a major attempt was made to soften this impact. The concrete panels on the outer walls use exposed aggregate derived from local stone, 'ground shaping' was used, access roads follow contours wherever possible and there was extensive tree planting to extend existing Forestry Commission plantations. The question of whether or not conifer planting is appropriate on the scale presently found in Britain has itself become highly controversial in recent years, but at Trawsfynydd the scheme was designed by the landscape architect Sylvia Crowe and there is much use of birch, rowan, gorse and heather.

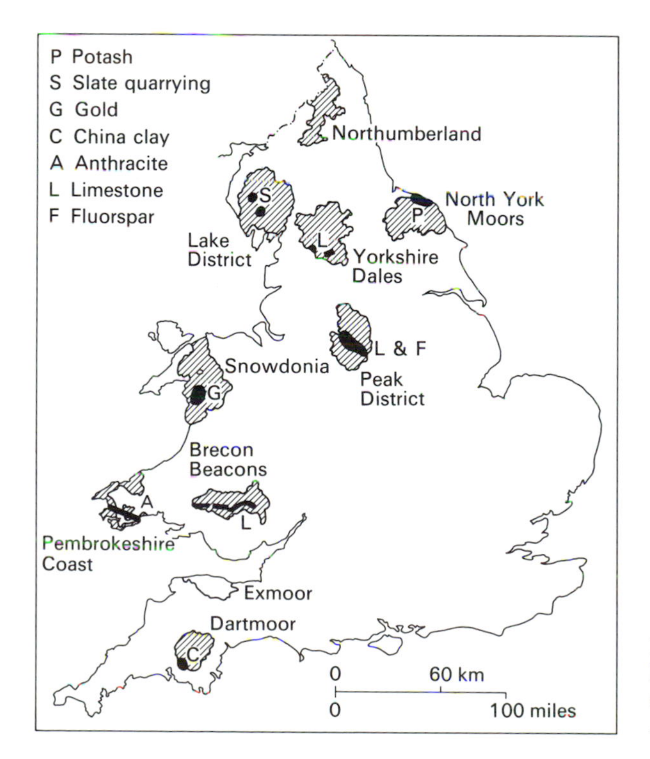

Mineral extraction in the National Parks. (Source: based on Countryside Commission, *National Parks Today*, 12.)

out onto the roads, but limestone is a vital construction material and provides much needed local employment. The Peak District (photo 68) is badly affected by this conflict but planning powers have been used with some success to curtail expansion of one or two quarries causing severe environmental impact. An application for extension of Eldon Hill quarry was refused by the Secretary of State in 1985. The extension would not meet the 'Silkin Test', that the limestone be required in the national interest and unavailable elsewhere. This line was again upheld in the same year with refusal for extension of Topley Pike quarry near Buxton. The Government subsequently issued a statement that appears to confirm the policy and should have the effect of discouraging further applications from the quarrying companies and shorten public inquiries, which all tends to support the idea that in certain specific cases conflicts can be 'resolved'.

Unfortunately, conflicts over the use of countryside are now being fought on a completely different scale. The deepest contemporary conflict involves a head-on clash between the interests of agriculture and forestry on the one hand and recreation and conservation on the other (see also chapters 1 and 2). From the early 1970s it began to be argued, by conservationists at least, that the countryside – the *whole* resource – was undergoing astonishingly rapid and irreversible change. The conflict was about comprehensive and extensive changes in land use brought about by agricultural intensification. As the conservationists saw it the changes were thoroughly deleterious affecting the beauty of landscape, flora and fauna that undermined the special character that had evolved over centuries. By the 1970s agricultural production driven by subsidies was seen as the main threat to the countryside and for the first time public attention became sharply focussed on the difficulty of dealing with this conflict when it was realised how separate and divided was the institutional and legal machinery intended both to safeguard and make use of the countryside. Indeed, many would still bypass that issue and argue that this current major conflict is particularly distinctive, but it is better viewed as the culmination of pressures that have long been fermenting. The conflicts that attracted earlier attention are not now irrelevant nor have they been superseded. There remain great problems over access to the countryside and a specific issue of principle, the question of military use (see photo 69) of some of our wildest and most beautiful places (including huge tracts in our National Parks), that has been simmering for some time looks set to erupt into a major public debate. What is important is that we should not continue with piecemeal strategies that seek simply to absorb each new conflict as it arises. The nature and depth of some contemporary conflicts call into question the framework of management itself.

68 *opposite* Tunstead, Derbyshire. Here the biggest limestone quarry in Western Europe – the main face is over a mile long – sprawls along Great Rocks Dale near Buxton right in the heart of the 'White Peak'. The quarry is amidst some of the best scenery and walking country in the Peak National Park, but when the Peak Park Planning Board was set up this quarry, together with several others nearby, was purposely left outside its administrative area so it can be said that most of the operations scarring the landscape in the photograph are not taking place in a National Park – a legal subtlety that is no doubt wasted on visitors who stumble on the quarry for the first time!

CONCLUSION

There seems little doubt that demand for countryside recreation will continue to grow. In future a more sophisticated public will want to explore much more of the countryside and use it for a much wider range of recreation activities. Unfortunately, experience shows that there is considerable scope for conflict over countryside recreation resources and as demand grows so must the role and importance of management, well resourced, positive management with new powers and a stronger organisational base.

From about 1950 there was extraordinary growth in countryside recreation, but through management it was possible to contain, even relieve, the most immediately pressing problems and conflicts. Critics might argue that recreation managers have been fighting a rearguard action through reactive policies that simply make a virtue out of necessity. The Countryside Commission's decision to promote footpaths, for example, goes along with a strong swell of pressure from the 'access lobby' and may be a case in point. But the opportunism in this action is quite understandable – the Commission has had so much to do, yet it is an organisation with few resources of its own. Its powers are narrowly drawn and, most importantly, it realises that in the short term it can achieve more by adopting a cooperative and conciliatory stance in the promotion of initiatives

that command strong public support. On the practical side of day-to-day management backed and encouraged by the Commission, great progress has been made and many policies have been successful. A more troubling criticism is that through stimulation of demand such policies have a ratchet effect on problems and conflicts (which incidentally secures the role of recreation management), but the fault here is much more fundamental and lies mainly with the Commission's paradoxical remit requiring it at the same time to conserve the countryside and promote its enjoyment by the public. Even so there is good reason to feel concerned about the social consequences of such policies. While physical consequences of increased demand often meet an effective practical response, social consequences are poorly researched and not well understood. There is little point in extra provision if it results in the levelling down of users' overall experience and satisfaction and decline in resource quality – the countryside is precious and the main aim must be deliberately to allocate resources to their highest potential for providing recreational opportunities and satisfaction.

However, it is in the broader sense of countryside management that there is a real problem. Although the explosive growth in recreation demand coincided with a huge expansion of the agriculture and forestry industries the resulting conflict between agriculture, forestry, nature conservation, recreation and amenity interests has only become prominent on the political agenda quite recently. The underlying problem is that too much has been expected from the statutory planning system. For the reasons considered nature conservation became separated from recreation management. In the circumstances it was natural for countryside managers to make the most of the framework they were given and concentrate on the conflicts and problems thereby rendered 'solvable'. Developments affecting the countryside, for example, were readily tackled through the radical, but negative, planning controls introduced from the late 1940s onwards and in that sense the protection of countryside stands as a most remarkable and enduring success. But the current major conflict shows that the way that other provisions to deal with recreation and amenity problems through the National Parks, other local authorities and the Countryside Commission came to hang on to the coat-tails of the main planning Acts was less suitable. It increased the separation from nature conservation and made for a managerial impossibility to confront agricultural change while there were forces driving expansion and ever greater intensity.

Of course, it is not the managerial problem but new political issues that have put this particular conflict on the agenda. A generation ago agriculture was enjoying considerable political and economic ascendancy, and farmers could change the landscape (mainly in lowland Britain) with a more or less free hand (and enormous public subsidies) and could easily resist new public demands for recreation. It is ironical that we have now come full circle and with problems of over production the concept of surplus land is being recognised and also perhaps that there are new opportunities for recreation and nature conservation. Is there also an opportunity for a more integrated and positive approach to management? That really depends on the political process, the new circumstances in themselves will not overhaul the political institutional framework through which management must operate. The Town and Country Planning Acts extended a large measure

69 *opposite* Worbarrow Bay, Dorset. The Dorset coast in this photograph fronts the Ministry of Defence's 3,000ha tank training ranges at Lulworth. The area, requisitioned in 1943 includes Tyneham village, centre of the main valley in the photograph, from which the residents were removed with a promise that they could come home after the War. Their return has never been possible and there was no public access for many years but the valley is now an interesting time capsule because the military presence has held the landscape in the past. There is a rich flora and fauna that in other places have been destroyed by modern farming practice. Since 1980 there has been a closely managed scheme to permit public access. Range walks are open most weekends, for all of August and for two weeks around Christmas. This network of paths gives access to spectacular coastal walks, the two ridges above Worbarrow Bay and Tyneham itself, and the main walk along the cliffs in this photograph is part of the Dorset Coast long-distance footpath.

of public control over private land, but compared with 40 years ago the situation is very different and to think of a very much broader extension of control over private land, if that is required for a more integrated framework of positive management, looks unduly optimistic. So we come back to the power struggle at the heart of the whole issue – who owns the land and who shall decide how it is used? That is what will determine the character of the countryside in the more leisured society of the future.

FURTHER READING

G.J. Ashworth, *Recreation and Tourism*, London, 1984.

J. Blunden and N. Curry, *The Changing Countryside*, London, 1985.

A.J. Patmore, *Recreation and Resources: Leisure Patterns and Leisure Places*, Oxford, 1983.

5 Industry

Manufacturing industry provides just over 40 per cent of the United Kingdom's export earnings and around 5 million workers, one quarter of the working population, depend upon manufacturing industry for their livelihood. While the continued importance of manufacturing industry to Britain's economy cannot be denied, its role has changed dramatically over the last 20 years and especially over the last decade. For the first time, in 1983, the UK imported more manufactured goods than it exported and from 1966 to 1987 employment in manufacturing declined from 9 million workers to 5.1 million, a decrease of 43 per cent. Around three quarters of a million of these jobs were lost from 1981 to 1987 and over the same period the proportion of employees in manufacturing jobs fell from 31 per cent to 26 per cent.

Britain's shift from a net exporter to a net importer of manufactured goods and the fall in manufacturing's share of total employment are each associated with the de-industrialisation of the British economy. An understanding of this process of de-industrialisation is critical to an interpretation of the changes taking place in the industrial geography of Britain. The introduction to this chapter examines the forces leading to de-industrialisation and then describes the organisations which influence industrial activity with particular reference to the roles of multinational firms and central government.

The fall in employment in manufacturing activity in Britain reflects both long-term trends and fluctuations associated with changes in the British and world economy. Clearly, Britain as the first industrial nation could not maintain its position for ever and some loss of manufacturing activity was almost inevitable as other nations developed their own manufacturing capacity. Increasingly, too, different nations have come to specialise in the different manufacturing industries in which they have competitive advantages and this too has led to further job losses in Britain. The situation has been exacerbated by low productivity arising from persistently low levels of investment in comparison with Japan and other European countries. Where investment took place it was not concentrated in those manufacturing industries in which output growth was likely to occur. The decline of manufacturing could also reflect the low status of manufacturing activity in Britain society. It can be argued that the 'best' people have not sought employment in manufacturing and the educational system is not sufficiently focussed upon industry's need.

These long-term trends have been accentuated by medium-term factors. Demand at home has grown but increased spending has taken place not on British goods but on imports. Many increases in demand have only a small spin-off on British manufacturing industry. For example, imports of motor vehicles and parts took 52 per cent of the UK market in 1984 compared with 20 per cent in 1972.

Britain also performs badly in terms of non-price competitiveness (for example, delivery time, reliability, design and after-sales service). Price competitiveness has worked against Britain too, the main element in the medium term being labour costs. Wage costs rose from 1974 to 1984 and these rises were not accompanied by sufficiently large changes in productivity to offset their effects. As a result unit labour costs in the UK grew more rapidly than in France, West Germany and the USA.

These long- and medium-term factors were supplemented by short-term influences which led to the marked fall in manufacturing output and employment after 1979. These were partly a result of the oil price rise of 1979–80. Although Britain was a net exporter of oil the recession which hit the world economy in 1980–81 reduced demand in Britain's export market. The Conservative Government's financial strategy from 1979 pushed up interest rates and further reduced domestic demand. The exchange rate rose to a peak in 1980, competitiveness of British industry slumped and there was a fall in manufacturing capacity of almost 20 per cent.

The long-, medium- and short-term factors all combined to produce the striking fall in employment in Britain's manufacturing industry but the overall employment decline hides some quite marked contrasts between industries. For example, employment in metal manufacturing in the first half of the 1980s fell by a quarter and employment in the manufacture of motor vehicles and motor accessories fell by one-fifth. In contrast, employment in instrument engineering rose by 3 per cent. A fall in an industry's employment does not necessarily reflect a fall in the demand for its products. Loss of employment may arise from changed working practices and investment in technical change to increase labour productivity. Where productivity is increased without an increase in sales labour must be lost and the greater the increases in productivity the greater the loss of labour. Most industries have lost labour both through increasing productivity and through loss of markets although the relative importance of each process varies from industry to industry. However, the significance of productivity gains is illustrated by the fact that, despite the much smaller workforce, manufacturing output in 1987 was 3 per cent higher than in 1979.

These changes arise from decisions made by the industrial organisations which manage and control Britain's industrial activities. The vast majority of British industry is run by the private sector. Among the private sector firms, over 30 years of financial restructuring has seen the share of net output controlled by the 100 largest firms rise from 27 per cent in 1953 to 41 per cent in 1983. The data, if anything, understate the role of large firms, for many small and medium-sized firms (the exact number is not known) are tied to them as buyers and/or sellers. The large firms are all multiregional and multinational ones.

Multinational firms are of two types: foreign multinationals with investments in manufacturing capacity in Britain and domestic multinationals with investments in manufacturing capacity overseas. There is no definitive listing of domestic multinationals nor is it possible to describe the geography of the employment they control but about 14 per cent of private sector manufacturing employment is in plants owned by foreign multinationals. Most of these firms are based in the United States (Ford, General Motors, Heinz, IBM) but there are investors from France (Peugeot–Talbot, Michelin), Germany (Hoechst) and the Netherlands (Philips). The percentage of regional employment in foreign-owned plants is highest in the South-East and East Anglia (20 per cent and 19 per cent respectively) and in Wales (19 per cent), Northern Ireland and Scotland (16 per cent each). The concentration of US-owned electronics plants in central Scotland has led some commentators to describe the area as 'Silicon Glen'.

Although large multinational organisations (whether based in Britain or overseas) make many of the decisions which influence the geography of manufacturing employment there is considerable debate over the role they play. To some they are the 'prime-movers' or driving forces behind the changes which take place, while to others they facilitate and speed responses to change but do not drive the system. A third viewpoint is that the changing geography of industry arises from the responses of firms to changes in economic forces and the role of individual large firms is of little significance.

While there is debate over the part played by individual firms there is no doubt that many changes in the geography of industrial activity are influenced by the activities of government. For many years governments have operated regional and urban policies – the role of these policies is discussed below – but other policies have spatial impacts too, notably defence policy. Nowhere is this seen more clearly than in the Bristol region and indeed this region is probably the most defence dependent region in Britain. As a result of the concentration of defence-related industries in this area it has received a major inflow of public spending. To set this spending in perspective from 1974/5 to 1977/8 the British Government spent £1,500 million on regional assistance, whereas it spent £890 million on defence procurement in the South-West region alone, a figure exceeded only by the North-West and South-East. Defence procurement spending in the south-east (£3,700 million) was more than twice the spending on regional assistance to the whole of Britain. The spatial impacts of defence policy decisions were emphasised in 1986 by the award of the contract for airborne early warning system to Boeing rather than GEC. Jobs may be moved or created in the plants of Boeing's UK sub-contractors (such as Plessey and Racal) but these may be accompanied by employment losses in GEC factories. In the geography of industrial change, which is the main theme of this chapter, the interaction of the state and the firm plays an important role.

INDUSTRIAL DECLINE

Three distinct but inter-related changes have taken place in the industrial geography of Britain in the recent past. There have been shifts of industrial activity:

1. From the inner areas of towns to the outer areas of towns. The share of an urban area's manufacturing activity in its central areas has declined and the share in its peripheral areas has increased.

2. From large urban areas to smaller urban centres and rural regions. The share of national manufacturing in large urban areas has fallen while that of the small urban centres and rural regions has risen.

3. From the north of Britain to the south. While the former area saw its share

70 *opposite* Inner-city industry, the Lower Don Valley, Sheffield, South Yorkshire. Sheffield is a city with a population of about half-a-million and a greater than average dependence on manufacturing industry. Much of its industrial activity is concentrated in the Don Valley and the section shown here lies immediately to the north of Sheffield city centre and within Sheffield's inner urban area. The area to the east of the main north to south railway line is typical of inner-city industrial areas in the mid to late 1970s. There is an extremely high density of industrial building with little space for any extension to existing factory buildings. Just to the north of an east to west railway in the south-west corner there is evidence of competition for urban land on the edge of the Central Business District where a motor showroom has encroached on the industrial area. The massive industrial buildings to the west of the main railway reflect Sheffield's links with metal manufacturing and the metal-using industries. Since this photograph was taken many of these buildings have been demolished and, indeed, in the top centre a brewery distribution depot is under constructions on the site of a former steel works.

of national manufacturing activity fall, the latter area saw its share of national manufacturing activity rise.

These shifts in industrial activity rarely involve the movement of a factory from one area to another, instead they are primarily a reflection of different rates of employment growth and decline.

It might be thought that these changes simply reflect the industrial mix of different areas. Common sense suggests that areas with industries whose employment is declining slowly or growing (for example, electronic and instrument engineering) will see their shares of manufacturing employment increase and those areas with a predominance of industries whose employment is declining rapidly (for example, motor vehicle manufacture and metal manufacture) will see their shares of manufacturing employment fall. However, industrial structure plays only a very small part in creating the three geographical patterns set out above, its influence being most obvious in the North to South shift.

Admittedly structure may play an important role in local economies dominated by a few industries but for most areas the key forces are the national trends described in the last section and the more localised influences which act upon industrial activity to create a better or worse employment record than the national average for that industry. These localised influences are explored by examining each of the three changing geographies.

Inner and outer cities

In the 1970s employment trends in a number of major British cities indicated a rapid employment decline in the inner city. Yet in some cities, for example Glasgow, more recent data show that decline has been fastest in the periphery. This raises the question of whether 1970s patterns are being replicated in the 1980s. Certainly one might expect inner-city decline to 'bottom out'. In many inner areas industrial decline has been so dramatic and only activities that are especially suited to central locations survive and the remaining parts of inner areas are characterised by extensive tracts of derelict land.

The rapid decline of industrial employment in inner city areas has attracted considerable public attention but it is important to stress that much of this decline is due to the increasingly poor performance of industrial employment in cities as a whole and the greater part of it results from the fall in the national numbers employed in manufacturing industry (photo 70). Nevertheless, the differential decline of the inner city compared with the outer city may reflect its historic role within a city, past planning policies, the characteristics of inner-city plants and, most importantly, the limited availability of certain factors of production in inner-city areas. Each of these explanations of inner-city decline can be considered in turn.

The growth in the population of a city is almost always accompanied by an outward movement of population and housing. Accessibility to an urban labour force in the past necessitated a location where population densities were high (the inner areas) or where, from the late nineteenth century, transport routes provided public transport facilities to all parts of the city (the inner areas). With

the wider dispersal of population in the twentieth century and with the increased personal mobility offered by both public and private motorised transport, industry has been able to develop outside the inner areas. Inevitably, the relative importance of inner city areas as centres of manufacturing activity has declined. Although, in aggregate, the inner cities share of manufacturing activity is falling, certain industries still have a propensity for inner city locations. For example, in Clydeside from 1958 to 1968 instrument engineering, textiles, clothing and paper and printing were among the industries which showed no tendency to shift their locational emphases towards outer-city locations.[1] Similarly, an analysis of inner city plants in Manchester which changed locations between 1966 and 1972 showed that no less than two-thirds of them chose to remain within the inner city.[2]

A policy measure which might have contributed to the more rapid decline in the inner cities manufacturing employment is the concentration in these areas of comprehensive redevelopment schemes, involving the clearance of areas of mixed housing and industry and their replacement mainly by public-sector housing. Although these schemes affected large areas in Britain's inner cities compulsory purchase and demolition of industrial premises played only a small part in the inner-city decline. In Manchester, only 9 per cent of the inner-city plants which closed were served with compulsory purchase orders.[3] There may have been other firms which were discouraged from operating in the inner areas because of the possibility of receiving a compulsory purchase order but the scale of this more subtle effect cannot be measured.

71 *opposite* Suburban industry, Beeston, Nottinghamshire, view from north. With a population of just over one-quarter of a million Nottingham, like Sheffield, had one-third of its resident population employed in manufacturing industry in 1981. The inter-war period saw major industrial development in the Trent valley to the west of Nottingham where the major employer in this suburban area is Boots, the pharmaceutical company. The company originated in Nottingham in the late nineteenth century. Initially its manufacturing operations were developed in an area to the south-east of the city centre and consisted of a jumble of old factories on the edge of some of Nottingham's worst slums. The suburban site at Beeston offered space both for expansion and improved working conditions. The first block was opened in 1928 and the second, described by Stanley Chapman as a 'shining palace of industry' opened in 1933.

Inner cities are characterised by young, small firms. These two features may well provide a small contribution towards the differences in employment performance between inner and outer areas since both new firms and small plants have a high failure/closure rate. New firms and small plants have been characteristic of inner-city areas for a long time but they do appear to have added a particular element of vulnerability in the recent recession. In Clydeside, for example, as the recession deepened after 1978 the rate of employment loss increased most dramatically in plants with between 11 and 20 employees. The extent to which large firms control plants in inner areas may also be important. This is because large firms have the flexibility to respond quickly to any cost disadvantage or advantages which arise from a plant's inner city location.

Cost disadvantages arise in a particular location from changes in the availability and price of capital, labour, and land and from changes in accessibility to markets, to supplies of materials and to services. For the most part the availability and cost of capital and labour is unlikely to vary within a city. Admittedly capital may not be readily available to some ethnic groups and thus developments in areas in which those groups are located will be restricted. Similarly some individuals may be 'trapped' in certain parts of the city because of their family commitments, low wages and/or part-time jobs which make long journeys to work unacceptable and these pockets of labour may be attractive to firms seeking low-cost non-unionised labour.

Accessibility to supplies and markets may have more influence than accessibility to capital and labour. In the past inner cities offered an optimum location for access to an urban market (that is at the centre of an approximately circular

urban area) and for access to suppliers of materials who also occupied central locations. The former is well exemplified by the central location of many old breweries while the latter is seen, for example, in the nineteenth-century gun and jewellery quarters in Birmingham. Today many firms serve regional and national markets and outer city locations offer easier access to inter-regional transport networks (photo 71). For those firms serving a specific urban market it can be easier to serve the city as a whole via ring roads thereby avoiding the necessity to move within or across the congested central areas. In terms of supplies, particularly of materials, most firms draw from national suppliers via the motorway system and inner-city concentrations of linked industries are now less important.

Increasingly there is evidence to suggest that the key factor associated with inner-city decline (if the most important national influences are excluded) is related to the nature of the costs associated with land and buildings in urban areas. This is also one of the factors associated with the urban–rural shift and it is explored fully below.

The urban–rural shift

The main features of the urban–rural shift are that London lost half its manufacturing employment between 1960 and 1981; free-standing cities like Sheffield, Cardiff and Edinburgh lost a quarter of their manufacturing employment, but rural areas experienced an increase of almost 25 per cent. For the period 1960 to 1981 there is a consistent pattern – the more urbanised an area the greater is its rate of loss of manufacturing employment (photo 72). By the late 1970s there is some evidence to suggest that the rapid decline in urban areas, like that of the inner cities, may be bottoming out. From 1978 to 1981 the conurbations (excluding London) had the highest rate of employment decline and the rural areas the lowest but London's employment decline was approximately at the same rate as that found in small towns.

72 *opposite* Urban employment decline, Halifax, West Yorkshire, view from east. From 1960 to 1981 manufacturing employment in large towns as a whole fell by 18 per cent, but in Halifax, an urban area with a population of 77,000, 36 per cent of the manufacturing employment was lost over the same period. As in many towns Halifax's industrial firms occupy constrained locations, the constraints being accentuated by the location of some of the major textile plants in the steep-sided valley of the Hebble Brook which crosses the area from east to west. The decline in manufacturing in Halifax was accelerated by the town's dependence on the textile industry. This industry, which lost half its employees nationally between 1971 and 1981, accounted for one-third of the town's employees and one half of the town's manufacturing employment in 1966. Dean Clough Mills in the centre of the illustration once employed 5,000 people; on closure in 1982 it employed only 600. The buildings are being developed by the private sector to offer managed workspace to small firms and in 1986 100 firms employing 500 people occupied the old carpet factory, now called the Dean Clough Industrial Park.

The causes of the urban–rural shift are complex. Certainly over the last two decades improvements in the transport infrastructure have made some, if not all, rural areas more accessible and a greater degree of accessibility may have been a permissive factor encouraging manufacturing employment growth in rural areas. The stimulus for such growth must lie elsewhere. A number of stimuli have been suggested and these include the influence of lower rural production costs. For example, it is often claimed that high rates work to the disadvantage of urban areas but a recent study has demonstrated convincingly that the rates levied by local authorities do not affect the location of manufacturing employment.[4] It may be too that manufacturing employment growth reflects the residentially attractive nature of rural areas. This may affect the location of entirely new plants and possibly influence the expansion of plants in high technology industries which use a pleasant environment to attract highly skilled workers. However, the influence of residentially attractive environments on the urban–rural shift is relatively small. There is some stronger evidence to indicate that central government policies to create new and expanded towns did affect the distribution of manufacturing employment along the urban–rural continuum in the south-east but its influence elsewhere was probably negligible. A major role for industrial structure has already been discounted so it seems that the major influences upon the urban–rural shift are spatial variations in the characteristics of the land and labour markets. Which is the more important is currently the subject of lively debate.

The influence of the land market or, more specifically, land availability reflects the operation of two forces. First, over the last two decades the number of workers per unit of factory floor space has steadily declined as firms were spurred to cut their costs by introducing newer and less labour-intensive technologies. Overall, despite declining employment firms have required more and more space and from 1964–81 the stock of industrial floor-space in England and Wales rose steadily. Second, availability of land in urban areas was constrained by both the high density of industrial buildings and competition from alternative uses so additions to floor space and jobs took place in rural areas (photo 73). In 1982, 53 per cent of the establishments in the cities in the East Midlands had between 80 and 100 per cent of their sites covered by buildings whereas in the rural areas only 27 per cent of the plants had this characteristic. Putting the two forces together leads to the argument that:

> the decline in employment density occurs on most existing factory floorspace and leads to large job losses in all areas while variations in the supply of land mean that increases in the stock of floorspace are concentrated in small towns and rural areas. In cities manufacturing employment falls because there

is little new factory building to offset job losses on existing floorspace; in small towns and rural areas the same loss of jobs on existing floorspace occurs but it is offset by the jobs created in new factories and factory expansions.[5]

Even if cities and towns offered high profit/low cost locations for industrial activity the limited availability of land would restrict new developments. These restrictions might be accentuated in those cities where green belt policy has limited industrial development on the city periphery. The characteristics of urban land which make urban areas unattractive to some industrial activities are seen most strikingly in the inner areas of Britain's cities. In the inner cities are found some of the highest densities of industrial building and for those parts of the inner city on the margins of the Central Business District there can be intense competition for land from, for example, retail warehouses, new shopping developments and offices.

Convincing as this argument may sound, does it really fit the 1980s? Many urban firms have redundant floor space in their premises, numerous old buildings are available to let and, as is shown below, promoters of industrial regeneration are only too willing to provide new factory units. If this is not enough there are often tracts of derelict land offering potential sites for development although the costs of reclamation may be very high. However, the general availability of land is not as important as the availability of land next to the existing factory. In the East Midlands only 30 per cent of city factories had vacant land adjacent to them but this was true of 46 per cent of the rural factories.

Increased demands for factory floor space mirrored competitive pressures encouraging firms to introduce new technologies to cut their costs. These competitive pressures and technologies have also led to the 'de-skilling' of some sections of the work-force. Among large firms a change of location offers the opportunity to introduce a new technology with a new work-force thus avoiding resistance from the old (an example is provided by the move of News International's London operations from Fleet Street to Wapping); even using the same technologies, a new location may give a new labour force with attributes more amenable to management. It has been claimed that, 'time and time again through the historical development of industry whole sectors have shifted location to escape a well-organised workforce'.[6] These arguments have led some commentators to see the urban–rural shift as a result of large firms running away from Unions. These firms seek out a malleable, non-unionised work-force, usually a female work-force with little tradition of factory employment. Such workers are particularly characteristic of rural areas, seaside resorts and smaller towns. Admittedly many of these small towns are in industrialised areas but by selecting female workers the unionised male work-force is avoided. Much of this argument depends upon stereotyped views of unions (anti-change) and firms (anti-union) and the evidence is mainly anecdotal. A brewery in central Birmingham is closed because of 'industrial relations problems' but equally the closure could have been interpreted as the firm favouring its plant in the smaller town of Burton which might have been more readily adaptable for new technologies. The extent to which smaller firms are able to benefit from assumed better labour relations in rural areas is difficult to judge. They might gain from the ability to adopt more flexible labour practices than their urban counterparts. The extent to which the urban–rural shift reflects a running away from the unions is likely to remain a matter of debate.

Whatever the causes of the urban–rural shift the second half of the 1970s and the 1980s saw attempts by both central and local government to restore the prosperity of Britain's urban areas. The problems of the cities were brought home to central government by inner-city riots and a number of cities have had a range of social and economic policies targetted upon them. The most dramatic have been the Urban Development Corporations (initially set up in Merseyside and in London's Docklands and now extended to some other urban areas) but central government has also stimulated Enterprise Zones (photo 74) and Industrial Improvement Areas (offering grants for refurbishing industrial premises and the environment in which they are set). Paralleling the policies of the national government, local authorities have set up Enterprise Boards (to invest in local industry), have begun their own programmes of factory construction, have encouraged the

73 *opposite* Rural employment growth, Fakenham, Norfolk. The southern part of the illustration shows the main industrial area in Fakenham, a small town with a population of around 5,000 set in the midst of rural north Norfolk. The centre of the market town lies just to the west of the photograph. Manufacturing employment in the Fakenham area increased by 20 per cent between 1971 and 1981 especially in clothing, engineering and food production. Fakenham industrial estate (south-centre in the photo) opened in 1972 and by 1976 it provided jobs for 71 employees in manufacturing. A Norfolk County Council survey showed that by 1980 employment in manufacturing on the estate had more than doubled to 181 employees. The industrial estate has encroached upon agricultural land to the south-east of the town and part of the field adjacent to the estate was purchased in 1986. This is being developed by the local authority as serviced land upon which new factory units can be constructed and it provides a clear illustration of the availability of land for industrial expansion in small country towns.

74 *opposite* Enterprise Zone, Dudley, West Midlands, view from north-west. Enterprise Zones were introduced as part of the Conservative Government's economic policy in 1981 and the majority of these zones are in large urban areas which experienced rapid falls in manufacturing employment. These zones cover only small parts of each urban area but within the zones the incentives offered (for ten years from the date of the zones designation) include exemption from rates and development land tax, and a simplified local planning system. The Peartree Industrial Park shown in the photograph occupies part of the northern section of the Dudley enterprise zone the northwestern boundary of which is marked by the railway line.

75 Science Park, Cambridge, Cambridgeshire, view from south. Science Parks now play a key role in the development strategies of many local authorities. In its traditional form a science park is a collection of high technology companies or research institutes set in well-landscaped and attractive surroundings and often linked into the research activities of a University. Unlike many of its successors the Cambridge Science Park was developed as a private sector operation under the guidance of Trinity College, Cambridge. Construction of roads and services began in 1973 on the northern edge of Cambridge adjacent to the northern by-pass running east-to-west across the picture. The two lakes shown in the illustration are key elements in the sensitive landscaping of the site. Overlooking the larger lake in the south is the Trinity Centre offering social facilities for those employed in the Park. By September 1986 the Park included just over half-a-million square feet of floor space and provided employment for around 2,000 employees.

growth of worker co-operatives and the manufacture of socially-useful products and have developed numerous advice centres. An interesting recent development is the addition of Science Parks to the development policies of local governments (photo 75).

It is too early to judge the overall effect of local and national governments' urban policies upon the urban–rural shift in manufacturing employment. Claims by local authorities that they have created a certain number of jobs have to be treated very cautiously for the sums involved in these policies are very small when set against the sums spent by central government on regional policies to redress some of the marked spatial variations created by regional shifts in manufacturing employment.

Regional shifts

Perhaps the most familiar of the differences in the industrial geography of Britain is the contrast between the North and South. In general Britain can be divided into two major areas to the north and south of a Severn–Wash line (see diagram below). Regions lying mainly to the north of this line have lost manufacturing employment faster than the national average, while in the regions to the south

(a) Employment Trends 1979 to 1985 (Source: *Regional Trends 1986*, 92)
(b) Assisted Areas, 1978 and 1984 (Source: *Go for Growth*, Trade and Industry Special Report, 3 March 1978, 27; *British Business*, 30 November 1984, 533).

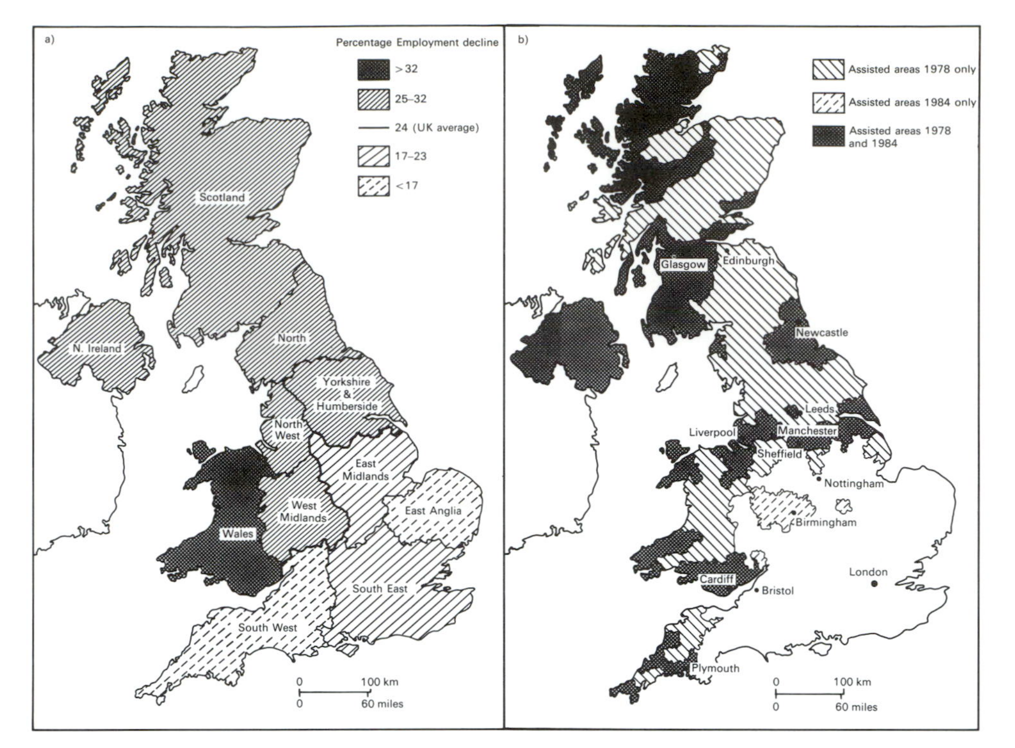

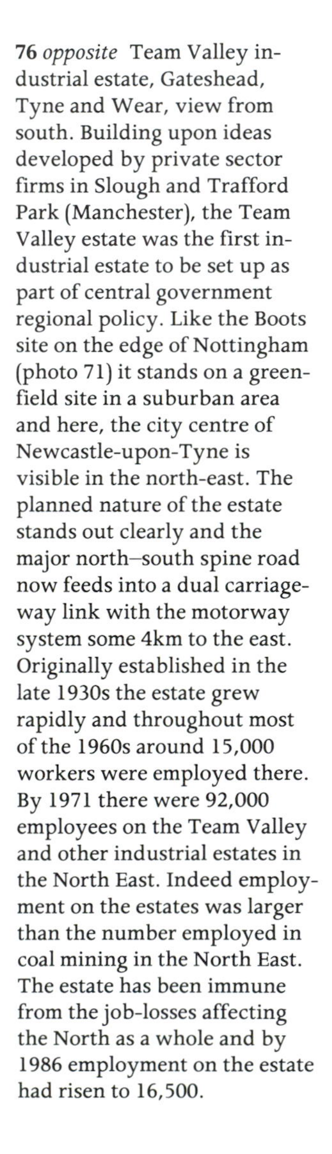

76 *opposite* Team Valley industrial estate, Gateshead, Tyne and Wear, view from south. Building upon ideas developed by private sector firms in Slough and Trafford Park (Manchester), the Team Valley estate was the first industrial estate to be set up as part of central government regional policy. Like the Boots site on the edge of Nottingham (photo 71) it stands on a greenfield site in a suburban area and here, the city centre of Newcastle-upon-Tyne is visible in the north-east. The planned nature of the estate stands out clearly and the major north–south spine road now feeds into a dual carriageway link with the motorway system some 4km to the east. Originally established in the late 1930s the estate grew rapidly and throughout most of the 1960s around 15,000 workers were employed there. By 1971 there were 92,000 employees on the Team Valley and other industrial estates in the North East. Indeed employment on the estates was larger than the number employed in coal mining in the North East. The estate has been immune from the job-losses affecting the North as a whole and by 1986 employment on the estate had risen to 16,500.

of it employment loss has been less than the national average. Whereas Wales lost 40 per cent of its manufacturing jobs between 1979 and 1985 East Anglia lost only 13 per cent. The contrast in the Midlands was particularly marked: whereas the East Midlands (astride the Severn–Wash line) lost employment at less than the national rate the West Midlands lost employment at the same high rate as Wales, Scotland and Northern Ireland. Of course, the exact position of this North–South divide varies over time and by early 1988 there was evidence to suggest that the line might be drawn from the Mersey to the Humber as both Wales and the West Midlands had begun, like the rest of the South, to lose manufacturing employment at less than the national rate.

One obvious interpretation of the North–South contrast is to suggest that regional performance reflects the mix of urban and rural areas within each region. Thus, it is argued the West Midlands performs badly because it is dominated by a conurbation whereas the East Midlands performs better because it is characterised by a number of large towns. However, detailed analysis shows that location in relation to the Severn–Wash divide is reflected in the performance of different types of area. Towns in the South, for example, have a better performance than towns in the North.[7] Clearly there is a differential regional performance which underlies that arising from the urban–rural shift.

The form of this regional shift has, at times, been very much influenced by regional policies implemented by central government. These policies have been reduced since 1979 but in the 1960s and 1970s regional policy played an important part in influencing regional differences although they were not able to change the fundamental contrast between North and South. In 1979 almost 50 per cent of Britain's working population was in areas eligible to receive some form of government aid. These Assisted Areas varied from those with very high levels of aid such as Clydeside, Tyneside, and Merseyside to those with much lower levels such as Humberside and South Yorkshire. The aid took the form of factory building together with grants and loans to manufacturing firms. The most visible outcome of the policy was the construction of industrial estates in the Assisted Areas (photo 76).

The overall impact of these policies was to create around 600,000 jobs in the most important of the Assisted Areas (mainly Scotland, Wales, the North, Merseyside, and Northern Ireland) from 1960 to 1981. It should be stressed that many of these jobs were not necessarily new to the British economy and that many

77 Plant closure 1, Invergordon, Highland, view from north. The Invergordon aluminium smelter was one of three smelters established in Britain in the late 1960s with strong encouragement from the Labour Government. It was believed that cheap electricity generated from nuclear power would enable British producers to compete effectively in the British and world markets. Although the government was advised that only two smelters would be viable three were built as a result of the strategies of the three firms involved. The Invergordon plant was constructed on the northern shore of Cromarty Firth which provided a sheltered deep water harbour for the import of materials (top centre), the site itself being adjacent to an existing railway for the distribution of aluminium ingots to rolling mills. The pylons and transformers associated with provision of electricity from the national grid are in the bottom left sector of the picture. A fall in demand for aluminium later necessitated closure of one of Britain's smelters and, after a relatively short life, the Invergordon operation was shut down.

78 Plant closure 2, Speke, Merseyside, view from east. The former British Leyland plant on Merseyside provides an indication of the contraction of manufacturing employment in the motor vehicle industry and the failure of a plant attracted to an area by regional policy. Set up in the 1960s in response to the Conservative Government's pressure to move employment opportunities from the West Midlands to areas of high unemployment it survived only about 20 years. Designed to assemble the Triumph TR7 sports car, the plant had a capacity of 100,000 vehicles per year. However, the car never sold more than 30,000 per year. Despite references to poor product quality and unsatisfactory industrial relations at the time of closure the real reason for the shutdown in 1978 was the continuing failure of British Leyland to hold its share of both UK and export markets.

of them were diverted from sites in the Midlands and the south-east. These were not necessarily jobs for all time and somewhere around a quarter of them had been lost by 1981 (photo 77 and 78) leaving a policy effect of around 450,000 jobs. As a result of the reduction in the number of areas eligible for regional aid between 1978 and 1984 and the subsequent abolition of a major element of regional assistance (the Regional Development Grant), the impact of regional policies in the 1980s is probably negligible and even its effects in the 1970s are open to debate. Some commentators can discern no policy effect after the mid 1970s while others argue policy continued to have a significant effect in the late 1970s.

Although regional policy may have damped down the contrasts between North and South these contrasts are of fundamental importance in the industrial geography of contemporary Britain. Whereas the urban–rural contrasts probably owe much to the ideas about constrained locations and the characteristics of their labour markets no such 'major' influences can be identified to provide an understanding of the regional shifts. It is unlikely that any major factor will be discovered and that what is happening is that a whole series of minor factors are all operating in a similar direction to favour the South. Their relative importance and the magnitude of their effects still await measurement.

Industrial structure is one of these factors. The south-east, East Anglia and the south-west from 1975 to 1981 had relatively favourable structures. For example, high technology employment growth from 1975 to 1981 was concentrated in the counties of Berkshire and Wiltshire (in the eastern section of the M4 corridor) and in other counties in London's western crescent (Herefordshire, Hampshire, Surrey).[8] Yorkshire and Humberside, Wales and the Northern region had particularly unfavourable structures. The Northern region had 18 per cent of its employment in industries whose employment fell by over 20 per cent (photo 79).

A second factor is that the South seems to be particularly favoured for new firm formation and product innovation in smaller firms. Empirical evidence has shown that new firm formation is encouraged in economies where there exist

a high proportion of small plants as is the case in the south-east. The reasons for this are not fully understood, but it has been suggested that small plants not only give individuals greater all-round experience than large plants but they also demonstrate to potential entrepreneurs that it is possible to go it alone. Innovative small firms tend to be characteristic of southern England too. Whereas 85 per cent of a sample of plants in south-east England had introduced a new product between 1973 and 1977 this was true of only 55 per cent of the plants located in those assisted areas with high levels of government aid.[10] This trend was not evident in the plants of large multinational firms which appear to spread the introduction of new and improved products throughout Britain's regional system, they do however concentrate their head-office and research activities in the south-east.

The large multinational firms which dominate the economy display a distinct spatial division of labour with an over-representation of white-collar jobs in the south of the country and an over-representation of manual jobs elsewhere. Nowhere is this seen more clearly than in the distribution of their office and research activities. The concentration of head-offices in south-east England is particularly striking. Of the hundred largest manufacturing firms, 88 have their head-offices in this region (diagram a) and 50 per cent of them are concentrated in just four London postal districts (diagram b). Estimates in the mid-1970s suggest that the head-offices of multiregional organisations contributed 50,000 jobs to an over-representation of white-collar jobs in the south-east, while offices administering divisions and subsidiaries of the multiregional organisations contributed a further 75,000 white-collar jobs to the level of over-representation.[11] The clustering of offices is explained partly by a desire to ease the development and operation of face-to-face contacts between organisations and the south-eastern location reflects the region's accessibility both to the rest of Britain and

79 *opposite* Structural influences, Consett, Durham, view from south. The closure of the integrated iron and steel making plant at Consett with a loss of 4,000 jobs in 1980 was a direct consequence of the rationalisation of steel-making capacity by British Steel in response to a fall in demand. The nine integrated iron and steel works in operation in 1978 had been reduced to five by 1984. Major job losses occurred both in surviving plants (Llanwern and Port Talbot in South Wales lost 11,000 jobs between 1980 and 1983) and through plant closures. Three of the plant closures resulted in the complete shut-down of steel making operations in the areas in which the plants were set. Shotton (in North Wales) was closed with a loss of 7,500 jobs and Corby (in the East Midlands) was closed with a loss of 5,500 jobs. The Consett plant was small and occupied an inland location, on the edge of the Durham coalfield. The open country stretching away to the north is clearly evident in the photograph. British Steel had unused capacity in its large and modern Teeside plant (70km to the south-east) which had direct access to a deep-water terminal for the import of raw materials.

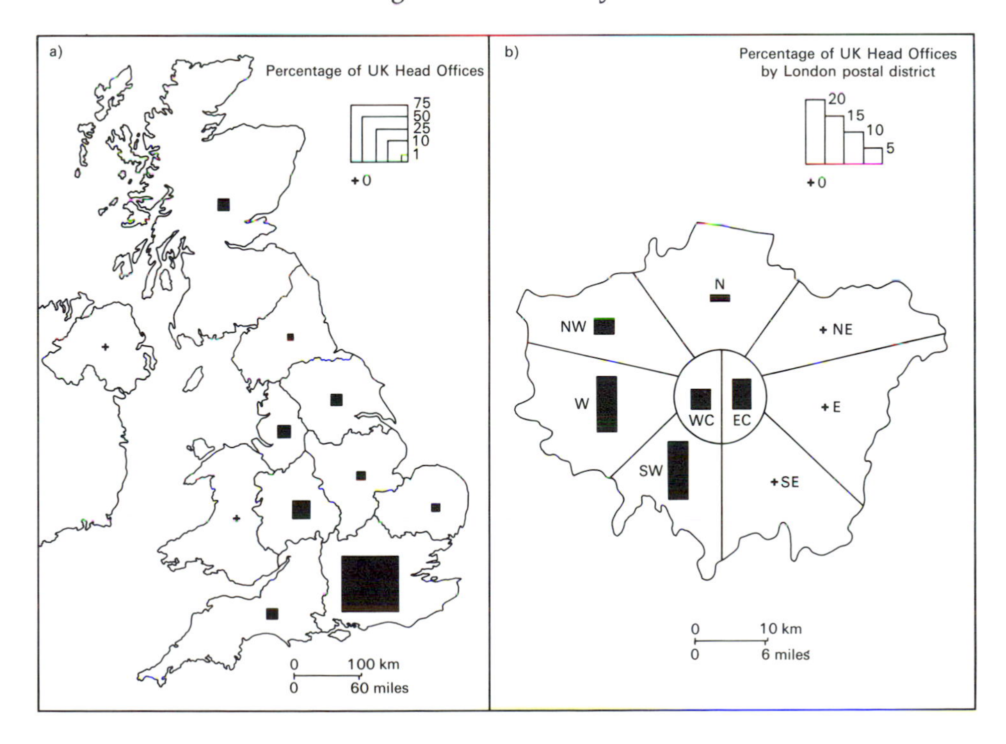

Regional distribution of the head offices of the 100 largest (turnover) manufacturing firms in the UK, 1982–3.
(a) Regional.
(b) London intra-metropolitan.
(Source: *The Times 1,000, 1982–3*)

80 Research and development, Hursley Park, Hampshire. Since 1958 Hursley Park has been the location of IBM's UK research laboratories and it is now one of IBM's largest research complexes outside the USA. Employing 1,600 people it provides the classic example of research and development facilities in the countryside of south-east England. Although occupied by one company its appearance has much in common with the Cambridge Science Park. Hursley village can be seen to the east while Hursley House (built between 1721 and 1724) occupies the north centre part of the photograph. At the time of the photograph IBM had built five major new buildings. In the last decade, another major building has been erected and car parking areas extended. This R&D site, like others of IBM, concentrates upon a particular area of research and its developments are used throughout the global IBM system.

the world, the concentration in the region of the major government and financial services and the fact that many of the firms which have come to dominate the economy have their origins in the south-east.

Although the research activities of these firms, like their head offices, tend to be focussed in the south-east they are in fact quite widely dispersed across a belt from East Anglia to the south-west. Just over 55 per cent of the research establishments of the private sector lie in this area of southern England. Within this belt R&D is associated with smaller urban centres and, in some cases, country houses set in rural environments (photo 80). Data on employment in research and development services of all types (not only those related to manufacturing) show that while less urbanised counties had 54 per cent of employment in R&D services they had only 23 per cent of employment in manufacturing.[12] In part the concentration in southern England reflects a desire for proximity to corporate head-offices, while the tendency to search for less urbanised areas probably mirrors the availability of sites and premises suitable for research activities and the attractive residential environments offered by the smaller towns and villages of south-east England. In some high technology industries accessibility to production units which are in southern Britain can also be important.

The concentration of head offices in the south-east has resulted in development of 'branch-plant' economies in other regions. By 1973, for example, 79 per cent of the employment in large plants in the northern region was in plants owned by firms based outside the region.[13] External control is usually assumed to lead to less commitment to a local area, limited linkages to the local economy and few white-collar workers. It is sometimes argued that externally-owned plants are more likely to close or run down employment than indigenous firms but the evidence for this is very conflicting and has shown no clear pattern.

This spatial division of labour is now widely recognised. Since white-collar jobs in manufacturing have declined at a slower rate than blue collar jobs their over-representation in the south-east may have contributed to the slower rate of manufacturing employment decline in the region. However the data on trends in white-collar jobs in manufacturing are difficult to interpret as firms are tending now to buy-in the white-collar-type activities from producer service firms.

Disadvantages for industrial operations in the North can also arise from the lack of producer services. These are the services such as computer bureaux which ease the operation particularly of firms which do not have access to in-house business services. Producer services which are growing tend to be concentrated in the south-east.[14] The provision of transport facilities also favours the South in that it has better international links, and is the focus of national transport services. Increasingly the North which used to be seen as peripheral in the British context is now also peripheral in the context of the European Community.

All these factors interact with one another to produce the differences in manufacturing employment change to the North and South of the Severn–Wash divide, and to these must also be added the most difficult of all to assess – the perceptions of decision-makers. The extent to which visions of a North of smokestacks and terraced housing contrasted to a South of rolling downs and cottages can affect overall patterns is difficult to judge but certainly advertisements for skilled jobs stress environmentally attractive production locations (photo 81). That jobs are being advertised in such locations demonstrates that despite the overall rate of job loss new jobs are being created.

INDUSTRIAL EXPANSION

The picture presented by the media of current manufacturing employment trends is one of unremitting employment decline and increasing industrial dereliction. In fact, jobs in manufacturing industry are being created in all parts of the country but they are hidden by the massive job losses which create the overall pattern of net job losses described in the previous sections of this chapter. The West Midlands and Coventry provide good examples of this process at work. From 1972 to 1975 no less than 86,000 new jobs were created in manufacturing industry in the West Midlands but this was not sufficient to replace the jobs lost through 145,000 redundancies over the same period.[15] Even in a period including the recent recession (from 1974 to 1982) some 7,000 jobs were created in Coventry.[16] Admittedly this is a small number when set against city-wide redundancies of 60,000 but the significant fact is that some job creation was taking place.

New jobs were created both by the establishment of new plants and the *in situ* expansion of existing plants. Significantly the rate of *in situ* expansion was

81 Residential attractiveness, Christchurch, Dorset, view from northwest. The southeast and southwest of England have strong links with high technology industry and it is sometimes argued that its development in this area reflects both the availability of workers with appropriate technical skills and the ability to attract such workers because of perceptions of the residentially attractive nature of southern England. The Plessey Defence System plant at Christchurch is a typical high technology plant located on the south coast. Plessey Defence Systems employ 2,000 staff, and 50,000 square feet of production floorspace manufacture command, control, communication and intelligence systems for the armed forces. Locational attributes play a major part in its recruitment advertising which stresses that 'sailing along the Dorset coast, horse-riding in the New Forest, beautiful scenery . . . [and] . . . thatched cottages . . . could be part of your everyday environment'. Could any analyst or analyst-programmer, resist such blandishments?

highest in East Anglia (one of the most rural regions) and lowest in the south-east (the most urbanised of regions) but regional variations are greatest in new plant openings and it is these openings which provide the most visible signs of industrial expansion.

An important group of new plants is those opened by entirely new firms. They tend to employ small numbers of workers in each plant but overall 98,000 jobs were recorded in enterprises new to manufacturing between 1966 and 1975. The regional pattern of these new firms is particularly interesting. The south-east, East Anglia and the East Midlands have a relatively high level of employment generation in new firms while the lowest rates (in the regions for which there are reliable data) are the northern and north-west regions.[17]

82 Greenfield site 1, Bridgend, Mid Glamorgan, view from east. Between 1976 and 1980 firms based in the United Kingdom set up 300 branch plants outside the regions in which they were already operating (regions are shown in the diagram on p. 146) and half these plants were located in the assisted areas of Wales, Scotland and the Northern Region; Wales alone receiving 30 per cent of them. It does not seem unreasonable to associate this inward movement with the availability of Regional Development Grants in these areas. One of the largest plants to be set up in Wales in this period was the Ford engine plant constructed on a greenfield site at Bridgend between 1977 and 1979 at a cost of £180 million. The large space demands of a modern factory are shown clearly. The plant is linked into the Ford production system via the rail link shown to the west of the plant. With over 1,000 jobs on offer Britain had to compete against other European countries and Wales had to compete against other UK areas to obtain the plant.

In addition to new firm formation, the establishment of branch plants by existing firms has been an important form of job creation. Most attention is usually focussed on the branch plants of foreign firms. The latest plant in this tradition is the Nissan plant opened in north-east England with 470 employees in 1986. From 1976 to 1980 about 100 plants were set up by foreign firms beginning production in Britain.[18] They created only 5,000 jobs in comparison with the 11,000 jobs created by such firms between 1972 and 1975. One third of the jobs in the 1976 to 1980 period went to Wales, while two-thirds went to Wales, Scotland and the Northern region taken together. The dominance of the assisted areas probably reflects the operation of regional policies, Department of Industry guidance, inducements from local authorities and the fact that these firms had no existing production base in Britain to influence their British location. Firms already operating in Britain (both UK and foreign-owned) have also located new facilities: one of the most notable cases is the Ford company's engine plant at Bridgend in Wales (photo 82).

New plants are found in almost all industrial sectors. Although new plants are typically associated with the expansion of demand for a firm's product they can also be associated with pressures to reduce costs to enable a firm to retain its market share. Faced with technical changes and competitive pressures multi-

site firms can select particular parts of their corporate production systems for new investment. In some cases an existing site is developed, in others a greenfield (new) location is selected for the new capacity. The brewing industry provides an example of the development of new sites in a long-established industry (photo 83).

Where new plants are located primarily to serve regional markets their location is strongly influenced to maximise accessibility to a particular regional market. In the influences on their location they have much in common with the location of manufacturers' warehouse/depot systems. Just as large manufacturing firms often have head offices and R&D laboratories separated from their production sites so stockholding sites may also be separated from production activities. This arises partly to minimise delivery times to customers but it is also encouraged by the extensive space needs of modern warehousing (photo 84). Just as new region-serving breweries are close to the motorway network so access to the motorway and dual-carriageway trunk road system is important for warehousing. It is perhaps significant that two of the three districts in Britain with above average growth in warehouse floor space from 1974 to 1982 are bisected by the M1. The importance of accessibility is stressed by the fact that not one district in Wales or south-west England experienced above average growth in warehouse floor space.[19] Of course not all warehouses are owned by manufacturers, some belong to firms specialising in distribution, while others are retail warehouses which are but a shade removed from the out-of-centre superstores and shopping centres described in the next chapter.

New warehouses and new plants create new jobs at particular places in the national economy but they may be linked (in either the short or long run) with job losses elsewhere. A new brick or cement works may result in the closure of older facilities and there may be a net loss of jobs to the firm and economy as a result of the new investment. In the longer run, new foreign-owned firms may eventually kill off the domestic opposition. A long-run scenario for the UK car industry might see the emergence of the northern region with the Nissan plant as a major centre for the assembly of motor vehicles and the associated closure of Austin-Rover operations in the West Midlands as the British firm finds itself unable to compete with British-built Japanese cars!

83 *opposite* Greenfield site 2, Magor, Gwent, view from east. The typical nineteenth-century brewery occupied a town-centre location and served the local urban market. Most modern breweries serve major regional markets and although some older breweries have been expanded, in a number of cases large new breweries have been built on greenfield sites in close proximity to motorways. The Magor plant occupies a 58-acre site adjacent to junction 23 on the M4 motorway. Representing an investment of £52 million it was opened in 1978 and has a capacity to brew over 1 million barrels a year. There is also sufficient potential to double the output of the brewery if this is ever required. The landscaped site with plenty of space for expansion and adjacent to a motorway is in marked contrast to the constrained location and congested road network associated with a town centre brewery.

CONCLUSION

The changes and trends discussed in this chapter created the industrial geography of Britain in the 1980s. By 1981 the small towns and rural areas were the location of virtually two-fifths of Britain's manufacturing employment. Nearly half Britain's manufacturing employment is in settlements with less than 100,000 inhabitants and south-west England now employs more in manufacturing than either the northern region or Wales. However, the south-east (with 1.5 million employees in manufacturing) is still the most important single region for manufacturing activity but it is the region which is least dependent on manufacturing. A less than average dependence is also characteristic of the two predominantly rural areas of the south-west and East Anglia and the two peripheral countries of Wales and Scotland. Overall, it is the central and northern regions of England that depend to some considerable extent on manufacturing activity.

It seems unlikely that the scale of the redistribution of industrial activity characteristic of the 1970s and the early 1980s will continue into the late 1980s and 1990s. There are signs that the differences in rates of manufacturing employment change between urban and rural areas are less marked and within the cities themselves the employment changes in manufacturing activity in the inner-city areas show less divergence from those of the outer city. In the near future the increasing significance of southern Britain may be offset by the operation of countermanding forces stimulated by the over-heating of the south-eastern regional economy. Traffic congestion on the M25 and high house prices are but two indications of this process at work.

Although emphasis has been placed through this chapter on the different employment performance of different parts of Britain (inner and outer city, urban and rural areas and North and South) the major contribution to decline in all these areas is the de-industrialisation of Britain as a whole. Nowhere is this demonstrated more dramatically than in the conurbations from 1978 to 1981 where manufacturing employment fell by 23 per cent and of this 19 per cent was due to trends in the economy as a whole and a mere 4 per cent to local factors. These national trends reflect the competitiveness of British industry in the international market place. The effects of this competitiveness on the national economy are the prime determinants of the degree of industrial expansion and dereliction in a local area. Unless the competitiveness of the national economy can be enhanced this country may become either a museum-Britain dependent upon such attractions as Edinburgh Castle and York Minster or a wilderness-Britain drawing tourists to the delights of Snowdon and the Yorkshire Dales.

84 *opposite* Distribution depot, Daventry, Northamptonshire, view from south. Examination of the growth of commercial and industrial floor space in England from 1967 to 1982 shows that warehousing was by far the most rapidly growing sector. By 1984 there were 128 million square metres of covered warehousing in England compared with 225 million square metres of industrial floor space. This warehouse (representing an investment of £30 million at 1972 prices) is the centre from which parts for Ford of Britain products are distributed to dealers in all parts of the world. The warehouse covers 1.6 million square feet, stands on a 138 acre site and employs 1,200 people. 400 tons of material are shipped in and out of the depot each day, 70 per cent of the output going to British dealers. A central England location on the outskirts of Daventry – midway between access points to the M45 and M1 motorways – offers a minimum transport cost location for a firm serving the national market from a single site.

FURTHER READING

M. Boddy, J. Lovering, K. Bassett, *Sunbelt City? A Study of Economic Change in Britain's M4 Growth Corridor*, Oxford, 1986.

S. Fothergill, M. Kitson and S. Monk, *Urban Industrial Change: the Causes of Urban Rural Contrasts in Manufacturing Employment Trends*, London, 1985.

W.F. Lever (ed.), *Industrial Change in the United Kingdom*, London, 1987.

W. Lever and C. Moore, *The City in Transition: Policies and Agencies for the Economic Regeneration of Clydeside*, Oxford, 1986.

B. Moore, J. Rhodes and P. Tyler, *The Effects of Government Regional Economic Policy*, London, 1986.

H.D. Watts, *Industrial Geography*, London, 1987.

6 Transport

At the turn of the century Hilaire Belloc formulated a law of minimum effort to describe why roads and other transport routes came into existence. His law aimed 'to find a formula of minimum expense in energy for communication between two given geographical points under given conditions of travel or carriage', and he then went on to explore the reasons why a straight-line route was not always possible. He later expanded these ideas to claim that the road has become one of the fundamental institutions of humankind as it controls the 'development of strategies', it provides the 'framework to all economic development' and it is the 'channel of all trade and, more importantly, of all ideas'. It should be noted that the book containing these views was funded by the British Reinforced Concrete Engineering Co Ltd, who may be seen as a predecessor of the road lobby.

During this century many of Belloc's dreams have been realised as individual mobility has increased, as living standards have improved and as technological innovation has influenced all our lives. Space has not remained constant but has decreased through time as improvements have taken place in the transport network. In Belloc's time the fastest train took two hours ten minutes from London to Birmingham, three and a half hours to Manchester and eight and a quarter hours to Edinburgh. These times have now been reduced to one and three-quarter hours, two and a half hours and just under five hours respectively. With the electrification of the East Coast main line the travel time from London to Edinburgh will be further reduced to about four hours. It may be quicker to route Glasgow trains via Edinburgh if the route between these two Scottish cities were also electrified (see diagram below). We are living in a shrinking world. International travel, even 50 years ago was slow and time consuming and involved only those interested in emigration, in trade, and in controlling the Empire. Now, the position has radically changed with over 8.5 million people taking package holidays overseas by sea or air, and a further 13.5 million people make international business, social and other trips. Transport has revolutionised the way in which we look at the world, our knowledge of the world and our concepts of travel (photo 85).

Transport has been a powerful agent in the agricultural, industrial and technological revolutions. Each respective revolution has been accompanied by rail, road and telecommunications development. The railways facilitated the industrial revolution with the switch from primary agricultural-based employment to a manufacturing-based economy. The car enabled the switch from manufacturing to service industry with the growth in service sector employment. Technology now permits a move to quaternary (knowledge-based industry) and quinary activities (home-services). Each of these transitions has involved an infrastructure

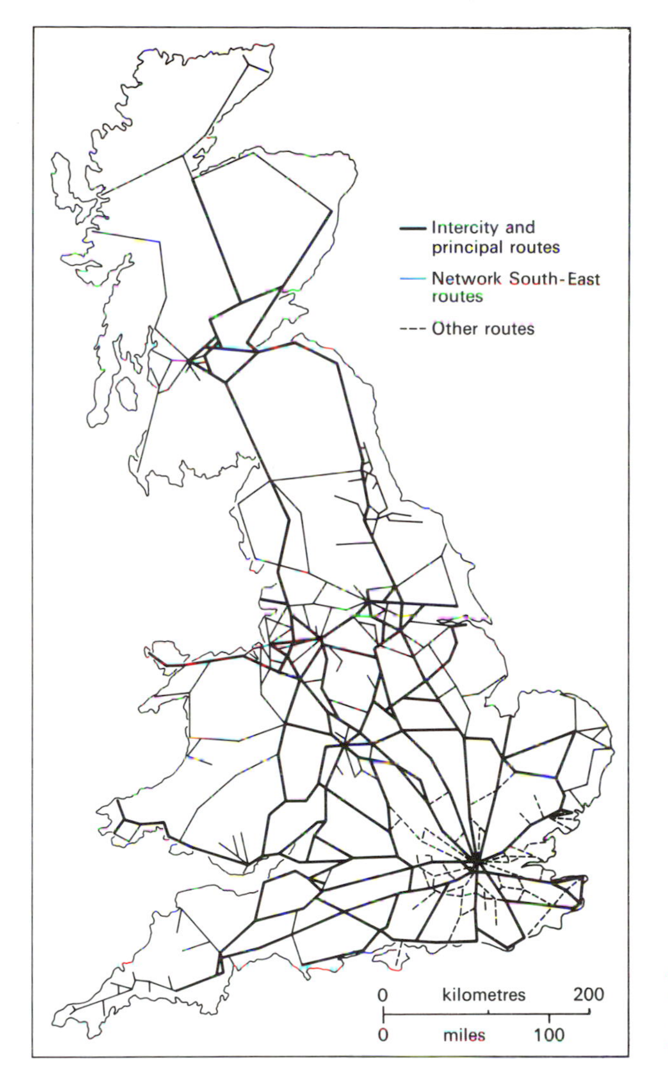

Great Britain rail network. (Source: based on British Railways, *BR Passenger Network*, 1989.)

and the development of a network – Belloc's law of minimum effort. The principal difference between the technological revolution and all previous revolutions is that it is aspatial and is not distance-related, even though it will have spatial impacts. Furthermore, the present revolution is knowledge-based and not energy-based.

Technological change will also have a major impact on lifestyles. The real debate is over the exact nature of the impacts and the differential effects they will have on particular groups in society. As Alvin Toffler has suggested in his influential book on the Third Wave, the transition is from an industrial and work-based society to one in which leisure pursuits dominate. These leisure activities will take place both in the home and at locations away from the home. Mobility levels are increasing, but the importance of the work trip is declining both in absolute terms (distance travelled per day) and in relative terms (proportion of total trips). Compensating increases are taking place in the shopping and social trip purposes. There is no explicit leisure category, but this omission may have to rectified in light of the changes in lifestyle outlined above.

Space may be shrinking, but so is time, as the rate of change seems to be accelerating. The railways took almost 80 years to reach their maximum extent and use, the car has really only taken about 35 years to reach its current level

of use, and most of the infrastructure has been built or renewed in that period (see diagram below). The rate of growth in technological innovation is phenomenal. For example, home computers were virtually unknown ten years ago. The one millionth was acquired in April 1983 and the current level of ownership is now over five million. Again, most of the telecommunications infrastructure has been modernised over that same period. These changes are consistent with others, namely the switch from public provision of services to private provision. For example, the switch from public to private transport, from cinema to the television and video, from the launderette to the washing machine, from the public to the private telephone and from the newspaper to teletext. Even though

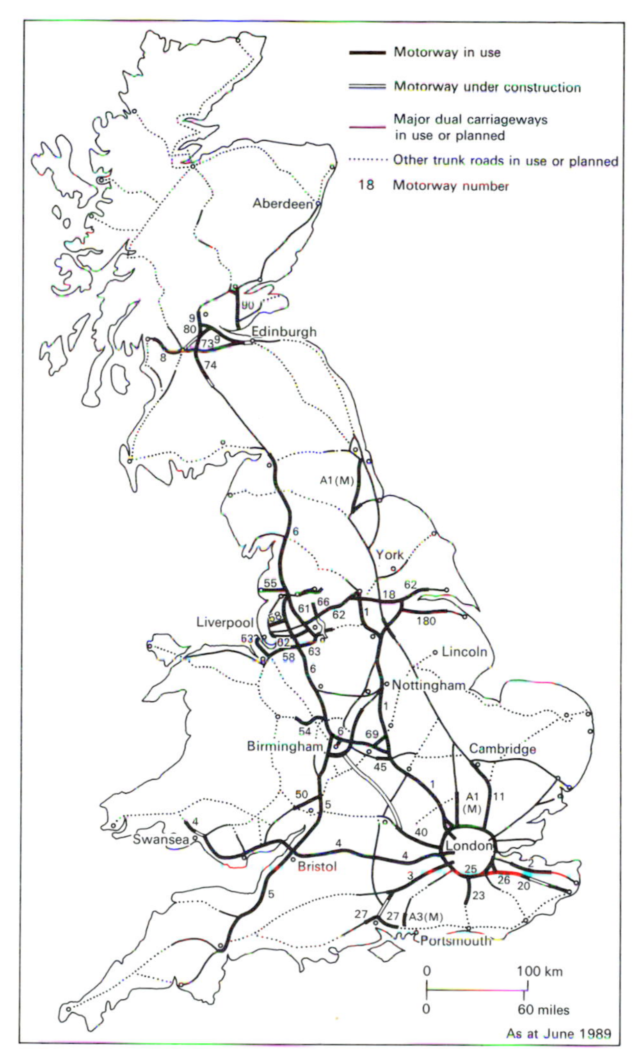

Great Britain main road network. (Source: based on Department of Transport, *Roads for Prosperity*, 1989.)

there is an increase in private ownership, the public sector still provides much of the essential infrastructure. But in each case the home has become the centre of activity. However, this switching is not available equally to all as not everyone can compete. The distribution is heavily in favour of the higher income households who are able to take advantage of new ideas. For the rest who are constrained through lack of resources or knowledge, the quality of the service may decline and prices may increase; these disadvantages are only partially ameliorated through subsidy to the service or the user. Polarisation of opportunity and use is one of the negative outcomes from affluence, the car and technology. The paradox is that with every revolution there have been increases in travel, not a decrease. The switch from public to private transport and to the use of information technology has increased mobility. However, the picture is more complex as averages conceal the distribution which has become more spread. The poor do not travel, and when they do it is by public transport or on foot and to local destinations. Potentially, everybody has much to gain, but in practice the poor are less likely to have access to innovation or the knowledge to use it.

85 *opposite* Heathrow Airport, London. Some 70 million passengers travelled by air to or from Britain in 1985 and over two thirds of these were on UK operators' flights. London's Heathrow airport is the world's busiest international airport, and handled over 24 million passengers in 1984 and over 0.5 million tonnes of freight. Gatwick, the second major airport in the London area, and the world's fourth busiest international airport, handled nearly 13 million passengers in 1984. The recently privatised British Airports Authority owns and manages the three main airports in south east England (Heathrow, Gatwick and Stansted), plus four airports in Scotland (Glasgow, Edinburgh, Prestwick and Aberdeen). Most of the other public airports are controlled by local authorities.

The £200 million fourth terminal at Heathrow was opened in 1986 (to the south of the picture) so that the capacity can be increased from 30 million to 38 million passengers a year. London Underground has built a 6km single loop railway to the new terminal at a cost of £23 million. At Gatwick a £100 million redevelopment programme has increased the airport's annual capacity to 16 million passengers. A second terminal will increase eventual capacity to 25 million passengers a year.

Views differ on the spatial impact of technological change. The late Peter Cowan believed that

> Many major inventions and innovations will have little effect on the overall pattern of settlements in Britain. For example, telephone, television, radio, videophones and many other technical advances are all essentially 'blanket type' inventions. There is no evidence to show that the telephone, by itself, has had any major influence upon the settlement pattern of Britain, yet it has been in use since the late nineteenth century.

The contrary view is stated by E.M. Forster in his short story 'The Machine Stops'. In this he describes a future society in which the movement of people from place to place has all but stopped. Individuals spend their whole lives in a controlled environment in which everything is brought from the central machine. Local travel never occurs and long-distance travel (by airship) is regarded as dangerous and eccentric. When the machine breaks down, society collapses. In retrospect, the impact of the various revolutions outlined above, including the latest technological revolution, seems to be the opposite to that described by Forster. While injustices abound, many people are now more mobile and less dependent on others than ever before.

In this chapter we focus on change in transport and the impact that this is likely to have (or has had) on life styles. We achieve this objective in two different ways. Firstly, we trace the changes that have taken place over the last 50 or so years with the invention of the motor car, its development and the effect that it has had on people's mobility. The review of the mechanical age is contrasted with a more speculative view of the changes which are likely to take place with the switch to a technological future based on increasing leisure and the use of telecommunications and information. Secondly, we examine a series of key political decisions which have been taken over the last ten years of Conservative Government. The concepts of the right to transport services, in particular public transport, have been changed with an extensive programme of privatisation, deregulation and reductions in the levels of public expenditure. Transport is seen very much as a commodity which can be bought in the market place at a market price. Intervention is only justified where the market is seen not to operate, and then only for good social reasons. These two changes are likely to revolutionise conventional wisdom as technology and government policy change attitudes to transport.

86 *opposite* Gravelly Hill Interchange, Birmingham. The Midlands Links motorway cuts through the urban area to the north of Birmingham and runs on a viaduct between Gravelly Hill and Castle Bromwich for 5.6km. Some 300 houses and 70 industrial premises were demolished in the course of construction. The most complex and spectacular intersection is at Gravelly Hill (shown here) where the M6 crosses the A38(M) Aston Expressway and local Birmingham roads. The interchange covers 12.5 hectares and includes 1km of the M6, 2.6km of the Aston Expressway plus 4km of interconnecting roads. The external costs of the motorway have been considerable with increases in noise levels and visual intrusion; but most controversy has risen over the long-term effects of lead in petrol on school children in the local area.

In the final section a choice is presented. Policies currently being pursued will lead to a mobility-based solution in which many people have unprecedented freedom to use their cars. However, there will be many people, perhaps even a majority, who will not have unlimited access to a car, and these individuals will probably be worse off. The alternative proposed is a future which provides accessibility for all. Technology offers the opportunity for such a future, but all the indicators seem to suggest quite the reverse. This is the dilemma for policy makers.

THE MECHANICAL AGE: THE CAR AND PERSONAL MOBILITY

A contemporary of Belloc was also anticipating fundamental changes in the twentieth century. Writing in 1902, H.G. Wells suggested that the great cities in Britain would 'diffuse' as motor traffic was segregated from other traffic on the roads. The motor carrier companies would compete in speed with the railways and cooperate with long-distance omnibus and hired carriage companies to form trunk routes. Even Wells did not envisage the speed of change brought about by Henry Ford's production-line methods introduced at his Highland Park works in 1913. The motor vehicle was no longer the sole domain of the public operator as it became an object which the private individual desired to own; this was also the beginning of the consumer age.

The car has altered the way in which we live and think. Over 60 per cent of households now have at least one car and that level continues to increase.

87 *opposite* Waverley Station, Edinburgh. Railways were pioneered in Britain and the Stockton and Darlington Railway (opened in 1825) was the first passenger public railway in the world to be worked by steam. Until recently, passenger journeys and freight traffic had been declining, but in the last few years a more positive image has been marketed, and this together with competitive pricing has resulted in traffic increases. The passenger network is divided into three parts – the Inter City network, the commuter services including Network South East, and the provincial services (see diagram on p. 161). British Railways has undertaken to reduce its Public Service Obligation (PSO – equivalent to a subsidy) by 25 per cent over the period 1983–87 and a further 25 per cent in the following three years.

As part of the deal, new investment (a package of nearly £1 billion at 1986 prices) has been taking place with the electrification of the East Coast Main Line, London to Harwich, and Glasgow to Ayr and Ardrossan; with new rolling stock and locomotives; and with new signalling and operating equipment. The electrification of the line between London and Peterborough was completed in May 1987 and it will reach Edinburgh Waverley Station in 1990.

To the user the car offers real advantages which alternative forms of transport can never match except in congested urban areas and over long distances. The car has a unique flexibility in that it is always available, it offers door to door transport, it has a high level of privacy and comfort, and it really forms an extension to the home. It is in effect a part of the home that can be detached and re-attached at will. These private benefits conceal considerable costs that car ownership and high levels of mobility have brought. As a result of the increase in car ownership the demand for public transport had fallen. This decline reflects both the direct transfer of passengers from bus and rail to the car, and the secondary effects such as increased fares and decline in service levels on public transport to compensate for the loss of patronage – the so-called vicious downward spiral. The car has also brought considerable environmental and social costs to society as a whole. The environmental problems of pollution, noise and vibration from transport are very apparent as are the personal and social costs caused by the 5,200 deaths on the road each year. Society seems to accept these costs, but the impact on those directly affected cannot be quantified (photo 86).

The car is also a significant user of resources. Car-based transport is the only sector in the economy that did not seem to react to the oil shortages of 1973 and 1979. Consumption of finite resources has continued to increase with many motorists either being insensitive to increases in petrol prices or being shielded from the direct impacts. The personal benefits of the car are seen to outweigh the public costs which include the consumption of finite fuels at an ever increasing rate. The car is only used for about an hour each day on average; for the rest of the time it is parked and occupying valuable space. Even when it is in use journey speeds are often slow as congestion reduces the efficiency of the road system and, in particular, influences the operation of public transport.

In addition one has to consider the basic inequality of the car. Not everyone has equal access to it and this will always be the case. Some will be too old or too young to drive, some will have some physical disability and some will not be able to afford to buy and run the car. There has been a polarisation between those who have unrestricted access to the car and those who are in some way disadvantaged. Public transport, taxis, the bicycle and other forms of unconventional transport, mainly in the voluntary sector, offer some compensation, but none provides the unique advantages of the car (photo 87).

Similar arguments can be made with respect to freight transport and the switch from rail to road and from public operators to private ones. The advantages to the individual firm of having its own fleet are again the flexibility and the control which can be exerted. Often, as with the car user, only marginal costs are considered as the fixed costs have already been paid. As with the dispersal of population, the dispersal of industry to peripheral and road-accessible sites has made rail less appropriate. Rail is best suited for long-distance bulk haulage of full train loads; over 90 per cent of rail freight is made up of this traffic. The position may change with the opening of the Channel Tunnel in 1993, as the European rail network will be directly linked to the British network without the usual costly interchange requirements (photo 88). However, the public costs imposed by a road-based freight industry are also considerable. Their impact individually, particularly for the largest vehicles, is greater than that for cars in environmental,

social and energy terms. These costs are likely to increase as heavier lorries are permitted on to the roads, even though the noise levels will be reduced.

Apart from the transport related impacts of cars and lorries, there have also been a series of structural changes which have been facilitated by private transport. Land use patterns have become more dispersed with the rise in the journey to work distances and acceptable average commuting times of 30 minutes in large cities to about an hour in London. Workplaces are no longer concentrated in the city centre but have become dispersed so that more travel is now orbital, not radial. Public transport provision becomes more problematical. Similarly, as workplaces have been dispersed, more travel is now made for other purposes, in particular for shopping and social purposes. With the trend towards concentration of shopping facilities in hypermarket developments and 'prestige' streets in the principal towns, one could suggest that this market is more appropriate for public transport operations; a similar comment could be made for health and social services travel as hospitals have also been redeveloped in larger and more specialised units. Public transport is ideally suited for the movement of large numbers of people to specific destinations (photos 89 and 90).

88 *opposite* Crewe Yards, Cheshire. The most important freight commodities are coal and coke, iron and steel, building materials and petroleum products. Over 90 per cent of rail-freight revenue is obtained from traffic in bulk commodities, most of it in trainloads. A network of 150 scheduled 'Speedlink' high-speed freight services have been established between the main industrial centres.

Gone are the days of the marshalling yards where trainloads had to be made up prior to dispatch (e.g. Crewe on this photograph). Most traffic now goes direct from source (e.g. a coal mine) to its final destination (e.g. a power station) in fixed train loads. Other traffic has been containerised. The dispatcher makes up a load then takes it by lorry to the container depot where it is transferred to rail for the trunk journey to a distribution depot for local delivery by road or to a port for export overseas.

THE POLITICS OF TRANSPORT

During the last 50 years, transport has remained outside the political arena. It was mainly seen as a technical exercise carried out by engineers according to a set of standards that had been evolved with experience. With the growth in car ownership congestion had increased and it was thought that the solution to the transport problem would be to build more roads; the maxim was to predict and provide. In 1963 the famous Buchanan Report changed the way people thought about traffic. The realisation was that a choice had to be made – either the city would have to adapt to the car or the car would have to adapt to the city. If the first alternative was adopted, then wholesale redevelopment would have to take place along the lines of the car-accessible cities of the USA. Alternatively, the car would have to be restrained in urban areas so that many of the historical centres could be preserved.

Having accepted the second alternative, other non-engineering issues become important. Evaluation of transport schemes becomes essential, and there was increasing public concern over social and environmental problems. Thinking was crystallised in the 1968 Transport Act which made a strong commitment to public transport with the return to deficit financing for the railways, and a new system of planning and financing bus services in the metropolitan and non-metropolitan counties. It was during the early 1970s that the effects of the rise in energy costs were felt with recession and increased unemployment. Inflation rose to double figures and public expenditure was cut back. The concern in transport was to make the best use of available resources; this was interpreted through extensive traffic management schemes and priority for public transport. It was during this time that many of the one-way systems were introduced, extensive pedestrianisation took place in some cities, urban traffic control was developed and bus only lanes became a familiar feature in many urban areas. The 1978 Transport Act went even further in its promotion of public transport. County councils now had a requirement to meet the transport needs of those people who did not have access to the private car. However, with the change of government in 1979, the generally low profile given to transport was changed. Whatever else might be said about ten years of Conservative Government, it has revolutionised many traditional views within transport, and has brought about the most radical changes in transport policy this century.

Previous governments had been content to introduce one major piece of transport legislation each decade. In the first half of the 1980s, the Conservative Government introduced six major pieces of transport legislation, one for each year from 1980 to 1985. This unprecedented activity is almost biblical in its pro-

89, 90 *this page and opposite* Devizes Locks, Wiltshire. The 29 locks are spread over 3.6km and have an overall rise of 72m; the record time for negotiating the locks was 2.5 hours. The side pounds are reservoirs for locking water to supply the locks. The locks are at present being renovated so that the entire length of the Kennet and Avon can be opened for pleasure craft. The canal was opened in 1810, extensively used until the Great Western Railway bought it in 1852, closed in 1948 and will be completely reopened in the early 1990s (120km in length). The views show the locks immediately after closure (1949; this page) and at present (1986; opposite) with all the pounds and some of the lock gates repaired. Note the course of the old railway to the South of the upgraded Devizes to Trowbridge road.

portions, particularly as the Government took a rest in 1986, the seventh year! At the centre of this revolution have been three primary aims:

—to transfer industries from the public sector to the private sector – privatisation or denationalisation.

—to introduce competition into public transport services and reduce the levels of public expenditure – deregulation.

—to encourage more private capital in major transport investment projects.

The Government has also been active in other sectors such as road safety, and the current review of road traffic law is likely to form the basis of a further major Act. Our attention will focus on the three issues above.

Privatisation

Transport has been the subject of much speculation about privatisation. The 1947 Transport Act introduced the concept of nationalisation as part of Herbert Morrison's plans for the reconstruction of British industry after the war. The British Transport Commission was set up 'to provide an efficient, adequate, economical and properly integrated system of public inland transport and port facilities within Great Britain'. Nationalisation covered British Railways, docks and harbours, and road freight. The Conservatives reversed the decision on road freight in 1953. The 1968 Transport Act established six nationalised industries – British

91, 92 *this page and opposite* The Isle of Dogs, London. London Docklands has been called the largest building site in Europe and tremendous change has taken place here in the last ten years. It covers a total area of 20sq km. There are four main areas – Wapping, The Isle of Dogs, The Royal Docks and The Surrey Docks. The existing population is 40,000 but this is expected to increase by more than 60 per cent in the next few years. One of the constraints on development has been the lack of adequate transport. Roads are being upgraded, but the major new investment is the Docklands Light Railway, opened in 1987. The railway, costing £77 million, links the Isle of Dogs Enterprise Zone with Tower Gateway (near Tower Hill) and with Stratford, with 16 stations on the 12km route. One 1.6km extension (costing over £90 million) will run from Tower Hill to a terminus at Bank, and a second extension may run to the East to Beckton. There is some concern over whether the railway has the capacity to handle estimated peak demand of 20,000 passengers an hour. In addition to the railway there is the STOLport (Short Take Off Landing) in the Royal Docks and a tremendous potential for opening up the river itself as a commuter route. The pictures show the Isle of Dogs looking downstream (1986; opposite) and upstream (1972; this page) – new buildings and the light railway viaduct can be clearly seen in the 1986 view.

Railways, British Waterways, British Transport Docks Board, National Freight Corporation, National Bus Company and Scottish Bus Group (photos 91 and 92).

Since 1979 the Conservative Government has returned some of these nationalised industries to the private sector, either in their entirety or through partial sales. The reasoning behind this policy is partly ideological and partly financial. The ideological argument is that organisations are more efficient in the private sector with the normal commercial pressures that competition brings. The financial argument is that the Exchequer can gain through the sale of companies, and that it can save through reductions in the levels of public support paid through subsidy or through lending to service capital debts.

The programme across the whole of the industrial sector has been impressive, but particularly so in transport. The most dramatic example has been the buyout of the National Freight Corporation by a management-and-employee consortium to continue the largest road haulage operation in Britain. The acquisition was the largest buyout ever undertaken in Britain and the first for a nationalised industry. In commercial terms this has been a remarkable success with some 20,000 of the employees (77 per cent) holding shares (amounting to 83 per cent of the equity) which increased in value from £1 to £47 during the six years 1982–8. In February 1989 the company was floated on the stock market with a value

SEACON

93, 94 *this page and opposite* Dover Eastern Docks, Kent. Dover is the most important gateway to mainland Europe and handles some 65 per cent of all cross-Channel car traffic and over 25 per cent of all roll-on roll-off traffic in the UK. New construction in the Eastern Docks has been necessary to increase capacity and access roads have been upgraded; at present nearly 14 million passengers use Dover each year and some 8.8 million tonnes of goods pass through the port annually.

Almost all the 27 million passengers using sea crossings do so on the cross-Channel or Irish routes. Cross-Channel hovercraft services take about a third of the time taken by ships and they account for over a quarter of all short sea traffic. The SRN Super 4 Hovercraft has been enlarged to accommodate 481 passengers and 55 cars – it is the largest hovercraft in the world. There has been a new access road built to the Eastern Docks and a considerable increase in container storage space has been provided as a result of land reclamation and the removal of the hoverport to the Western Docks.

of £950 million. The largest privatisation is the sale of British Airways. This company ceased to be a statutory body in 1984 when it was re-established as a Public Limited Company under Government ownership. In February 1987 the Government sold its shareholding in the company to private investors for £892 million (photos 93 and 94).

Deregulation

The second change has been the deregulation of the bus industry, the most important policy switch since regulation was first introduced in the 1930s to protect a vibrant growth industry against predatory practices. Deregulation as it affects all urban and non-urban areas outside London has been introduced in two stages. The 1980 Transport Act deregulated long-distance bus services (those with a minimum journey length of 48 kilometres), excluded small vehicles (less than eight seats) which were not being run as a business, and introduced the idea of trial areas where there would be no requirement for road service licences. The 1985 Transport Act has effectively created a trial area for the whole country. The commercial network only has to be registered, and the local authority has no control over the operator provided that he or she runs the services as registered. The subsidised network complements the commercial network with

95 Ribbleshead Viaduct, Yorkshire. The Settle-Carlisle railway extends for 115km and has 17 viaducts and 12 tunnels. It has recently been under continuous threat of closure and British Rail are now looking to the private sector for a purchaser to run the route. One reason for this is the repair cost of the Ribbleshead Viaduct with its 24 arches built up to 50m above the valley floor.

each service being put out to tender by the local authority which has to specify the route, levels of service and fares (in most cases). London has been given an intermediate position under the terms of the London Regional Transport Act (1984) with only limited and carefully controlled competition.

This two-tier system treats each route on its own merits. Again, the Government argues that competition will bring greater efficiency into the public transport industry and end the extensive practice of cross-subsidisation. Perhaps the main reason was increasing concern over the escalating costs of grants and subsidies to the bus industry which totalled over £1 billion in 1983, with a similar figure for the railways. Deregulation can be seen as a radical change in Government policy. Previously, some form of public transport provision was almost seen as a right, an essential part of one's quality of life, particularly for those with no car. Public transport provision is now seen very much as a market commodity, and it is only when the market fails that intervention should be considered and only then if there are good social reasons for action. The question here is whether transport services can be viewed within the market framework or not – only time will tell.

The corollary to deregulation is the question of subsidy to public transport, a problem which has tested researchers for many years. There are several points. First, should the subsidy be given to the particular user in the form of a concessionary fare or to the operator for running the service so that all users benefit? Secondly, how should the subsidy be allocated to the different forms of public transport? At present, about half the subsidy goes to bus services and half to rail services. However, it is the higher income groups which use the rail services and make more trips; the subsidy per trip on rail is much higher than that for a bus trip of a similar distance. In terms of value for money it would be better to put all the subsidy into improving bus services. If equity arguments or those of equality are used, then the same conclusion would be reached, namely that the disadvantaged make fewer trips than average and those they do make are by foot or short bus trips. The Serpell Committee (1983) estimated that the commercial rail network which might be sustained in the long run with no financial support from public funds would only include 16 per cent (2,600 route km) of the existing network. Such extreme measures would make the Beeching proposals for rationalisation seem positively benign! Perhaps the rail system should be divided into the commercial and subsidised routes with the latter being put out to tender; conversely, privatisation of the entire system could take place either on a national or a regional basis to re-establish the pre-1947 situation (photo 95).

The third subsidy question relates to the public and private sectors. The subsidies for public transport are well known and under constant review, but the subsidies to the private motorist are at least as great. Over 50 per cent of all new car registrations are in a company name and many organisations also pay for insurance, tax and petrol for their employees. Motorists who do not have to pay the true costs of their travel will use the car more frequently and make more trips. About a third of all vehicles on the road receive some form of company assistance. In addition to the direct personal benefits, there are also external costs to others; included here are environmental costs, the loss of patronage to public transport, and the congestion costs imposed on other road users. There seems to be a basic inconsistency in policy as to whether subsidy is necessary at all and, if so, whether it be restricted to public or private transport or both. Perhaps there is a case for deregulation in private as well as public transport; this would mean some form of road pricing and charging for the use of road space.

Finally, there is now considerable pressure on the EC to allow competition in European air fares. Most fares are fixed and complex market sharing arrangement have been worked out between operators. It seems unlikely that complete deregulation will take place in Europe as occurred in the USA (1978), but there may be some modest change. Operating costs are some 20 per cent higher in Europe than the USA, but average fares are 35 per cent higher.

Infrastructure

Britain completed its first generation of motorway construction with the opening of the last section of the M25 around London in 1986. Progress has been steady since the building of the Preston Bypass in 1958, and there are now some 2,800km of motorways. However, it seems that the 1990s may be a significant period for new investment, not just in the motorway network. Much of the urban infrastructure is Victorian in construction and needs extensive renovation, in particular the sewerage and water systems (photos 96 and 97). In the transport sector several large new projects have been planned and approved – the Channel Tunnel, London's third airport and the electrification of the East Coast main railway. The debate over whether we need more roads is likely to re-emerge with a second generation network. These major projects influence us all, both directly through shorter travel times and indirectly through the urban environment, employment opportunities or paying for them in taxes.

However, these impacts are not the same for all groups in society and their effects are often long term. The Government plays a series of important roles (fiscal, regulatory, advocacy), but crucially the objectives are not always clear and are often contradictory. We may need fast, high-quality roads, but do we need the increased levels of pollution, more consumption of energy and the loss

96, 97 *this page and opposite* M25, London. In 1969 the Greater London Development Plan proposed three Ringway roads for London. As a result of vociferous objections the Layfield Panel of Inquiry recommended that only a single complete orbital road should be built. The final section was opened in 1986 and the £1 billion 187km route passes mainly through the Green Belt. Already some of the congested Western sections are being upgraded from three to four lanes as they are carrying nearly 120,000 vehicles a day (42 per cent over design capacity), and it is planned to widen the whole of the M25 to four lanes in each direction. Despite delays the road offers quicker journey times for traffic using it and it has provided relief to communities along the route.

This page shows the M25 under construction near Havering (1981). This section was opened in 1983. Extensive screening has taken place along the entire motorway with an average of 20,000 trees per mile being planted; despite landscaping and some screening, noise and visual intrusion are still considerable. Shown opposite is the opened motorway and Bell Common Tunnel at the northern tip of Epping Forest (1986). The M25 has been placed in a 470m tunnel to reduce environmental impacts. The forest and a cricket pitch were restored on the 'roof' of the tunnel.

of agricultural land? It is here that the public inquiry has been a very effective mechanism for both delaying unpopular decisions and in some ways getting them abandoned altogether. Local opinion differs; some acknowledge that more investment is required to sustain economic growth; others are indifferent to the growth arguments but resent the local impacts of the project. The political risks posed by major projects are high and the financial commitments are now considered too great for politicians concerned with reductions in public expenditure. The solution being adopted involves the use of private capital and share issues to fund large-scale transport projects so that the private sector can take the risk. Similarly, Hybrid Bills are being introduced in Parliament and debate limited to the select committees. This procedure avoids the necessity for large, costly and lengthy public inquiries, but has been bitterly criticised by objectors.

It seems that projects will be planned and implemented at a much faster rate than in the past if private finance and a much-shortened inquiry procedure are used. Hence the second generation of infrastructure investment may take place in the 1990s. Examples include the proposal for a second bridge across the Severn and the new bridge to supplement the two Dartford tunnels. This second proposal has been put forward by the Dartford River Crossing Company (Trafalgar House) and will provide four new lanes to be used by southbound traffic on the M25. The Select Committee has now completed its inquiry and approval has been given for construction to begin. This private scheme will open in 1991. The problems of finance and the public inquiry do not arise and the government has taken no risk (photo 98).

98 *opposite* Forth Road Bridge, Lothian. Bridges and associated structures account for between 25 and 30 per cent of total motorway expenditure. It is not Government policy to charge for the use of motorways as their purpose is to relieve congestion on parallel routes. The only situation where charges will be made is where there are significant time savings and no direct alternative. This means estuary and river crossings such as the Forth Road Bridge, the Severn Bridge and the Dartford Tunnel. The Forth Road Bridge is not a motorway bridge but an essential link between the M90 to the North and the M8 and M9 to the South of the Forth, and it was opened as a toll bridge in 1964.

The most prestigious project is the Channel Tunnel which epitomises the new optimism with largescale transport infrastructure projects and the use of private capital. The Channel Tunnel Group – France Manche (CTG–FM) will receive no public funds or Government financial guarantees, and so has to take the risk itself. The Governments have given assurances to investors that there would be no political interference or cancellation, and that the promoter has full commercial freedom to determine policy (including fares levels). The original proposals were evaluated by a Joint Assessment Group of British and French experts in November and December 1985 where it was claimed that all issues were discussed, including engineering feasibility and design, capacity, travel times, psychological factors, safety of navigation, environment, economic effects and employment. A decision was made to reject three schemes; Channel Expressway's road and rail tunnel, Eurobridge's rail tunnel and road bridge, and Euroroute's rail tunnel and road bridge/tunnel. The accepted scheme was the most modest proposal for a rail-only tunnel. A Channel Link Treaty was signed in February 1986 as a commitment to the project and a Hybrid Bill was introduced in Parliament. No public inquiry was held, only Select Committee hearings where a tight schedule allowed only limited discussion of many local concerns, in particular about the site of the tunnel entrance and facilities at Cheriton just north of Folkestone. The Bill became law in 1987 and construction has begun on both sides of the Channel. Capital has been raised in the City and through a public share flotation, and the tunnel should open in 1993. Costs have escalated by over £2 billion and it seems likely that the developers will have to raise new finance through loans from the City and a further equity flotation (photo 99).

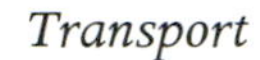

With respect to major infrastructure projects, the Channel Tunnel provides the procedural model for the future. The consultation procedures which have often delayed road proposals for up to ten years and the decision on London's third airport for nearly 20 years have now been short circuited through the Hybrid Bill. The Government now takes only limited risks as the capital has to be raised privately. The general public may feel that they have been bypassed and that they now have no say in decisions that often affect them directly. It may only be a matter of time before the same principles of private capital and short-cut consultation are applied to new road construction projects.

THE TECHNOLOGICAL AGE: TELECOMMUNICATIONS AND INFORMATION

We are now entering a period of change which is likely to be as significant as that caused by the advent of the car. The rapid changes brought about during the age of mobility and the move of transport into the political arena may both seem small compared with the potential initiated by the technological revolution.

The short term

The short-term impacts are already being felt in transport. In-vehicle communication through two-way radio and telephone contact is already well established and provides the traveller with information so that the car can be used as a mobile office. British Telecom's Cellnet and Racal's Vodafone both operate cellular phone systems and have over 578,000 users split about half and half between the two systems. The UK has the largest system in the world and there are plans to give 90 per cent of the population access to the system. Costs still remain high for installation and use, but as with all technical innovations, prices will fall. More advanced systems of in-vehicle communication could help drivers with route choice and prototypes have been demonstrated in Germany and Japan, and a research programme (Autoguide) is now under discussion in Britain.

Technology has also been used in monitoring and controlling vehicles in congested urban conditions. Urban traffic control systems which rely on historic data to determine the appropriate sequences for traffic lights are operating in more than 35 cities in Britain. Seven cities have now installed active systems which respond to actual situations as they occur, and road pricing has been successfully applied in Hong Kong. Automatic identification of priority vehicles (e.g. buses, ambulances and fire engines) can help in improving their speeds and in giving them priority at junctions. Automatically guided vehicles are a slightly more futuristic option. Various systems are available for bus guidance (e.g. the Daimler-Benz O-Bahn system in Essen), and personal rapid transit has already been introduced in several pilot schemes. Even though the technology is available, development costs are high and public acceptability is often low.

Many of these technological changes have already been introduced either through in-vehicle communications or through urban traffic control systems. The former makes travel by car more attractive and the latter allows the traffic engineer to squeeze more capacity out of a given road system. As such, the direct impact of each is likely to be similar, namely that more travel will be encouraged. The same transport infrastructure is having to accommodate significantly increased levels of traffic, and technology is now the principal means by which this increase is achieved. A second outcome of technological innovation in transport has been significant reductions in labour costs as the industry has become more capital intensive. Transport production lines have been automated and there have been reductions in the numbers of staff required to operate services. For example, one-person operated trains (e.g., on the London underground) could be replaced by completely automated services, and the number of crew in aeroplane cockpits could be reduced from three to two. The opportunity costs are high, particularly in the transport sector where labour has traditionally been highly unionised. However, the employers argue that significant increases in efficiency and productivity can only be achieved through reductions in labour costs.

The long term

Longer-term effects are harder to pinpoint. Transport may not cause change, but

99 *opposite* Channel Tunnel: the Cheriton Terminal Site, Kent. A link between England and France was first suggested by Napoleon in 1802. It has been started several times but never completed. The 1986 scheme is for two single-track tunnels (7.3m in diameter) plus a service tunnel (4.5m in diameter), and the total length will be about 50km from Cheriton (northwest of Folkestone) to Frethun (southwest of Calais). Its maximum capacity will be 4,000 road vehicles in each direction per hour, and the system will operate as a roll-on roll-off shuttle between the two terminal points noted above. Both passengers and freight will use the tunnel, and it is estimated that existing ferry journey times from London to Paris and Brussels will be more than halved to just over three hours; it will even give quicker city centre to city centre journey times than air. The whole scheme is to be privately financed and was initially expected to cost £2.33 billion (at 1985 prices). Since that time cost estimates have increased to £7 billion as contracts for rolling stock and construction are much higher than expected. The main debate is over the possible rail link between the tunnel and a new terminal in London. The decision has been delayed as there are questions over how the finance can be raised privately, the environmental costs of the link, the exact route to be followed, and whether the terminal will be at King's Cross or at Stratford in East London. The tunnel will be opened for use in 1993 and it is estimated that it will carry 42 per cent of passengers and 18 per cent of freight traffic across the Channel.

it has certainly facilitated change, and the car may be a prerequisite for transfer to a high technology society. Development pressures have taken place along transport routes. The 'Golden Triangle' to the west of London has three distinct axes, each with an airport, good rail and road links:

North via the M11 and Stansted to Cambridge
West via the M4 and Heathrow to Reading
South via the M23 and Gatwick to Crawley

Each of these axes has now been joined by the M25 and development pressures on the Green Belt are now apparent.

Technology has an important permissive role as it allows the home to become the centre of most activities and with the reduction in work travel it has increased the amount of leisure time available. Many people now work from home and it is possible to shop from home through the Telecard which uses a 'gateway' from Prestel to take orders directly and arrange delivery on the same day. Similarly, the Club 403 system in the West Midlands is aimed at families where there are two working adults for whom the weekly supermarket trip becomes difficult to fit in. Homelink services between the Nottingham Building Society and the Bank of Scotland allows direct access to banking facilities. The cashless society is with us and the move to direct debiting by electronic transfer is now becoming widespread. However, it is worth stressing again that not all individuals have equal access to the technology. Only in specific and limited situations (e.g. shopping schemes in Gateshead and Bradford) can a social concern be identified.

Conventional analysis argues that technology either substitutes for travel (e.g. through telecommunications or teletext) or there is some complementarity which results in the modification of travel patterns. Technology increases the efficiency of transport or transport increases the efficiency of technology. However the impact may be synergetic in the sense that increased use of technology causes *more* use of transport. There seems to be no evidence that any revolution has resulted in less transport. Lifestyles may change significantly but the net effect is that some people, even perhaps a majority, travel more. In spite of these apparent benefits, transport policy should be concerned with all individuals and society as a whole, not just those who can enjoy high levels of mobility.

TRANSPORT POLICY FOR ALL

It can be argued that successive governments have avoided a comprehensive statement on transport policy. The Labour Government attempted such a commitment in 1976 with Transport Policy – a Consultation Document, but their efforts were not implemented as there was a change of government shortly after the process of consultation was completed with the Transport Act 1978. The energy crisis, world recession, high inflation and unemployment all proved to be more important issues for government action. Although the last eight years have brought tremendous activity, we are no nearer to a transport policy except that the tenets of a market approach, efficiency and value for money seem dominant; this dominance is implicit rather than explicit.

Policy should be concerned with providing access for all and transport should facilitate participation in activities. There may be a move away from the high mobility alternative where the concern is with the quantity of travel towards a broader concern with the quality of life, and technology offers the opportunity for this switch. The car provides unprecedented freedom for those who enjoy the exclusive use of it, but it is never likely to be available to all people. Over the next few years the number of people who will not be able to use the car is likely to increase, not decrease, as the elderly and dependent population becomes greater. The role that government should play is to ensure that the policy objectives are followed, either through legal or fiscal controls. At present, it seems that governments take action, but nowhere is it clear how this fits in with the overall policy. This requirement should be made explicit and it should embrace both public sector and private sector operations.

In the urban area, such a policy can easily be applied so that the appropriate balance can be reached between public and private transport. It would involve restraint on the use of the car through road pricing or physical restraint (e.g. area licensing schemes) or effective parking control, or a combination of two or three restraint measures. As noted above, the car is only used for an average of about an hour a day. For the remainder it is occupying valuable space, and even when it is in use it is causing congestion. Restraint through pricing or control would have to have an impact on all users, so any form of company subsidy or parking space would have to be eliminated to ensure that all needs were competing equally. However, the most crucial element in any urban policy would be a high quality public transport system. In most urban areas this service would be provided by bus and minibus services together with taxis. Bus-only streets would be introduced where only public transport would operate. In addition, some of these vehicles could be adapted to allow the disabled access to the transport network. In the larger cities trams and light rail transport could be introduced along corridors of heavy demand. They could either run at high speed along separated routes or at low speeds along the street network. In the city centres priority would be given to cyclists and pedestrians with other vehicles operating at low speeds. Safe cycling is one form of transport available to most of the population, but in only three cities does cycling account for more than 18 per cent of journeys to work (Cambridge 27 per cent, York and Oxford 20 per cent each). High use of cycles requires flat terrain and a low level of perceived risk. The second condition is best achieved through a dedicated network of cycleways. Even where this is not possible, speed limits could be reduced to 30km/hr so that cycles can be ridden safely with other traffic. These proposals may seem revolutionary and impracticable but they are not. Many similar schemes have been introduced in European cities with startling results. Accessibility for all has been improved, use of cycles has increased, the environment has benefited in general and accidents have been reduced. It seems that the outcomes from such a policy would have positive scores on the transport policy objective.

In addition to transport-based policies, long-term land use decisions could also ensure that facilities are locally based, within cycling and walking distance. Technology could also help to improve access to these facilities without the need for any travel. The capacity of the urban road network has always provided a challenge to transport planners, but again technology may provide the answer.

It can help in squeezing more out of the system and it can allow for more flexibility in work hours and places of work. As society moves to the quaternary and quinary stages more work can be done at home and there is more scope for flexible working; the outcome may be an end to peak-travel demand and individuals choosing when and where to travel.

Fewer opportunities are provided for public transport outside urban areas as distances are greater, fewer people want to travel and destinations are more dispersed. The car has a much greater role to play in these situations because of its flexibility. But at the interface between town and country there is still an opportunity for transfer from private to public transport provided that interchanges are made attractive. Again the full costs of the car should be borne by the user: company subsidy should be removed and the tax element in petrol should fully reflect the external costs. A progressive tax would relate the cost of motoring to the use of the car. Similar arguments are relevant to the freight sector with progressive tax on fuel that reflected all costs and damage caused. Alternatives to the car in rural areas would include a wide range of options – coaches, buses, minibuses, community buses, dial-a-ride, postbuses, shared cars and taxis. Schemes would depend on local circumstances. Long-distance travel would also offer a range of alternatives to the car such as air, rail and coach. In each case the scope is considerable, but the same basic policy question would be raised, namely as to which alternatives provide accessibility for all.

Comment is required on financing and efficiency issues which have dogged various policy initiatives. A policy that is focussed on accessibility is concerned with the quality of transport and whether participation in activities can be undertaken locally. So the total consumption of resources should reduce and the environmental costs of transport should decrease. Similarly the polarisation between those who have a car and those who do not should be reduced as the policy is directed at making opportunities equally accessible for all people. It is likely that the costs of travel will increase for the car user in urban areas through the switch to progressive taxes on petrol and restraint on car use. In non-urban areas costs are not likely to rise for those car users who only travel an average amount; heavy car users would pay more. Company car subsidies would be eliminated so that costs would be imposed on all car users equally; similar increases would affect freight hauliers.

Public transport would be able to operate more efficiently in urban areas through exclusive routes and so costs would be saved. There may also be a case for transfer of subsidy from rail to bus as the costs of bus support per kilometre are much less than those for rail. Bus is also more accessible for many people than rail services. Capital would be required for infrastructure investment in pedestrianisation, cycle routes and tram or light rail transit schemes. Overall however, the costs of a total policy package on the public purse are not considerable, particularly if the private sector can be involved as in the proposed rail transit systems in Manchester and Bristol.

With respect to efficiency, it is recognised that changes are required in the operation of many public transport organisations. The present Government has chosen to increase efficiency through competition and privatisation, but this is often at the expense of the user; this option reduces accessibility as service quality declines and prices rise. Competition in the true sense does not exist in public transport as there are too many restrictive practices. Competition has tended to lead to instability followed by oligopoly and the eventual re-establishment of monopoly. Short-term gains lead to long-term loss. The problem is to match up efficiency with accountability, and this can be achieved in both public and private sectors.

Routes can be franchised so that services have to operate at approved levels of fares. In the private sector this works in taxi operations where the service is provided at a known price. The drivers are mainly self-employed and use various means to maximise their income. In the public sector the system of franchising at fixed prices has been adopted in London with some success, and it seems that the halfway move to full deregulation gives greater flexibility to London Regional Transport than to operators elsewhere in Britain. It also offers better possibilities for service coordination and information to the public. Crucial to efficiency is the use of the most appropriate management and accounting practices. It is here that more research is required to determine whether worker control in private companies leads to greater efficiency or not.

The alternatives are clear. The minimum intervention approach is based on the greater use of the car, on much longer travel distances and the use of technology to facilitate this growth. However the high mobility option is resource intensive and increases the polarisation between those who have a car available and those who do not. The 'accessibility for all' argument proposed here examines the role of transport in the context of facilities and attempts to reduce travel distances and make more use of environmentally beneficial forms of transport which use less energy. Car users would be penalised so that all people benefit. Technology is used to reduce the need for transport and to spread demand across the whole day. If decisions are not taken and there is no clear statement on transport, then policy evolves in a haphazard manner without the implications of such a non-policy being followed through. The time has come for a fundamental review of transport policy and a clear statement of alternatives, otherwise we shall mindlessly enter a future determined by the car and by technology. This is the choice to be made.

FURTHER READING

J. Brotchie, P. Newton, P. Hall and P. Nijkamp (eds), *The Future of Urban Form*, London, 1985.

K.J. Button and D. Gillingwater, *Future Transport Policy*, London, 1986.

Council for Science and Society, *Access for All? Technology and Urban Movement*, London, 1986.

P. Hall and C. Hass-Klau, *Can Rail Save the City? The Impacts of Rail Rapid Transit and Pedestrianisation on British and German Cities*, Farnborough, 1985.

A. Harrison and J. Gretton (eds), *Transport UK – 1985: An Economic, Social and Policy Audit*, Newbury, 1985.

A. Harrison and J. Gretton (eds), *Transport UK – 1987: An Economic, Social and Policy Audit*, Newbury, 1987.

E. Wistrich, *The Politics of Transport*, London, 1983.

7 Energy

Britain does not have a comprehensive energy policy. There is, however, a set of goals, policies and programmes which broadly defines the conventional wisdom in the energy field. The closest approach to an all-embracing policy statement in recent years can be found in the 1978 Green Paper, a consultation document published by the then Labour Government.[1] It argued essentially that oil and gas would run out early in the twenty-first century, that demand for energy would continue to grow and that Britain's energy future depended on coal, conservation and nuclear power – the 'CoCoNuc' strategy. Although the energy situation has changed substantially since 1978, the basic tenets of the Green Paper have continued to dominate official thinking, albeit with much less emphasis on conservation than on expansion of supply.

One reason for the lack of a clear strategy is that the Conservative Administration, in office since 1979, has been anxious to be seen to interfere in the energy market as little as possible. But energy supply in Britain has, at least until recently, been largely under Government (or oligopolistic) control, so in spite of the rhetoric there is a considerable amount of state intervention. For example, exploration and depletion of North Sea reserves is influenced by Government allocation and taxation policies; the Coal Industry Act of 1980 sets financial targets; a nuclear power programme was set out in a White Paper in 1979 and an 'Energy Efficiency Office' exists under the auspices of the Department of Energy. Policy also evolves in the interstices of other political processes, for instance in evidence and cross-examination at major public inquiries, and in Government responses to reports from Commissions and Select Committees which have examined energy issues. 'Energy Policy' could be defined as the sum of these parts, shaped by history, politics and economics.

Britain has undergone a transition from a one-fuel to a four-fuel economy since the Second World War (see diagram below) three features of which have been particularly important in shaping the current energy situation. The first is the decline of the coal industry, with its attendant social costs. British coal met nearly 90 per cent of Britain's energy needs in 1950; it now provides 37 per cent. Until the early 1970s the major policy question for coal was how to mitigate the effects of this decline. Coal's 'second coming' which seemed to be presaged by the 1973 oil crisis has, so far, proved to be illusory. Attention now focuses on improving 'efficiency' in the industry and there are serious question marks over its future. The second major feature has been the growth in Britain's dependence on oil and gas. Throughout the 1950s and 1960s the world price of oil fell in real terms and Britain's imports steadily increased, the cheapness and convenience of oil overcoming nagging anxiety about security of supply. The economy was already

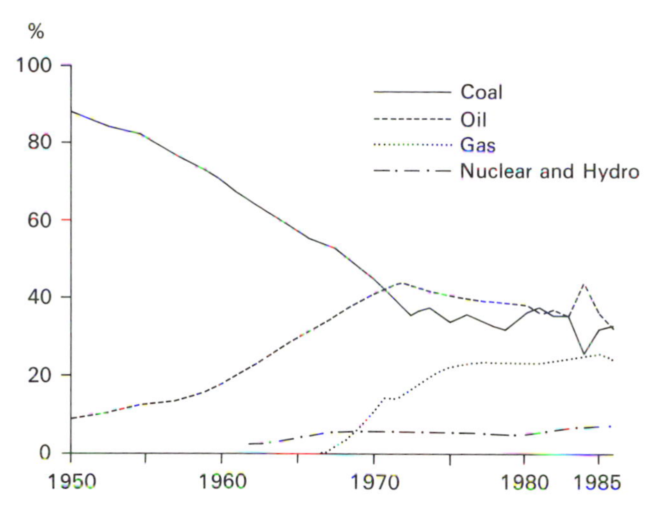

Energy supply 1950–86. (Source: *Digest of Energy Statistics*, 1987.)

geared to oil when indigenous resources were discovered in the North Sea, and Britain has become self-sufficient in net terms in the 1980s. The market share of natural gas has also expanded rapidly to meet almost a quarter of primary energy requirements. A third feature of the transition is the slower than expected growth in the contribution from nuclear power. Difficulties can be attributed to indecision over reactor choice, construction delays, cost escalation and, especially in the last decade, public hostility, which seems unlikely to decline in the aftermath of the Chernobyl disaster. After 30 years of development, nuclear power provides some 6 per cent of British primary energy. The 'bright promise of the benign atom' is proving difficult to realise and may yet turn out to have been a mirage. The current supply mix is unlikely to remain stable for long. The factors which will affect it and the ways in which policy should seek to influence them in coming decades are crucial and controversial issues.

The availability of resources must to some extent dictate energy policy, though it is not a one-way relationship since policies (relating to petroleum revenue tax, for example) also influence reserves estimates. Britain is rich in energy resources, and is currently self-sufficient in net terms. Oil and gas reserves are likely to last into the twenty-first century and coal reserves are sufficient for several hundred years at current rates of use. Britain has an indigenous nuclear industry, one of the world's best sites for a tidal barrage and significant potential (though little political will) to develop other renewable resources. Ironically, the very abundance of energy options has contributed to Britain's relative inertia in energy policy when compared with countries such as France, with its aggressive nuclear programme, or Denmark, with its vigorous commitment to energy conservation.

Certain basic tenets of energy policy have been so resilient in the post-war period, almost regardless of the government in power, that they constitute a firmly established 'conventional wisdom'. First, energy policy in Britain has been growth-orientated. According to the 1978 Green Paper, energy supplies must be 'adequate' and 'this implies that they should not be a constraint on economic growth'. Furthermore, it has hardly been questioned that projected demand for energy must be met; projected demand is thus equated with need and translated into energy programmes. Recently, however, interesting questions have been

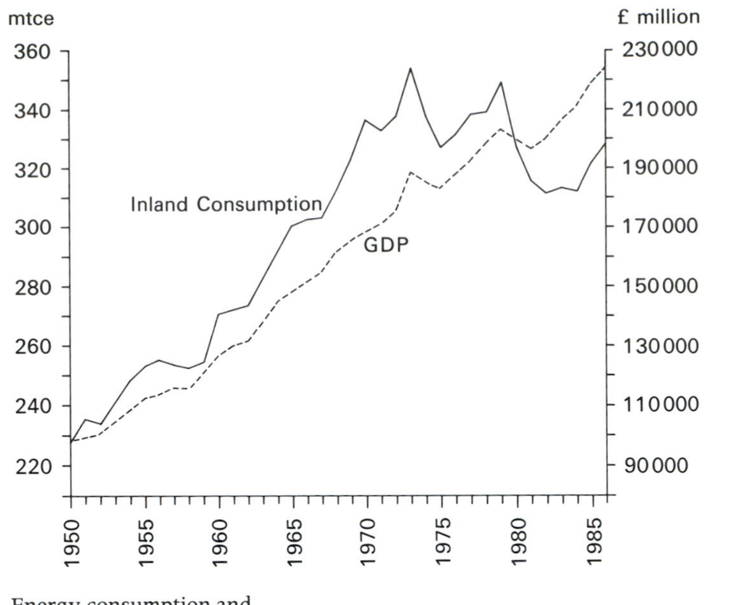

Energy consumption and economic growth 1950–86. (Source: *Digest of Energy Statistics*, 1987.)

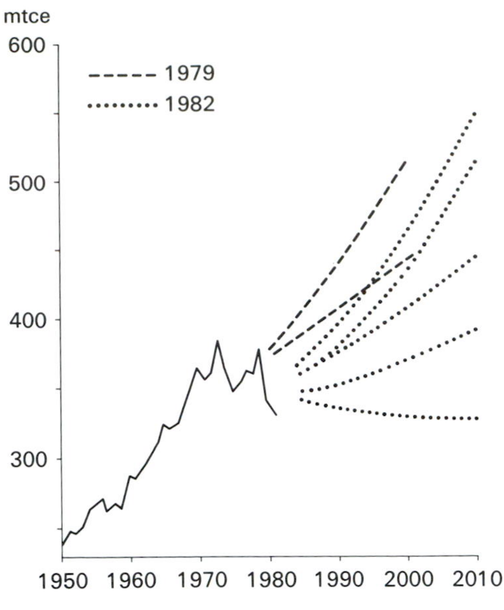

Projections of total primary energy demand based on differing scenarios. (Source: Department of Energy, *Evidence for Sizewell Inquiry.*)

raised about the relationship between economic development and energy consumption. Does energy demand grow in direct proportion to the growth of the economy? More radically, does energy consumption need to increase at all in a growing economy? The energy co-efficient – growth in energy consumption divided by growth in gross domestic product – used to be a well-behaved statistic (see diagram above), having approximated 0.7 throughout the period of steady economic growth in the 1950s and 1960s. Since the early 1970s, however, the relationship has become very unstable. Between 1970 and 1980, while the British economy grew by 18.2 per cent, energy consumption actually *decreased* by 2.7 per cent (an energy co-efficient of −0.15). There is no single explanation for this change; it arises from a combination of factors including higher energy prices, recession, conservation and industrial restructuring. Together with uncertainty about the economy and other factors, it makes forecasting based on the relationship between macro-economic indicators and energy demand a precarious exercise. Uncertainty is demonstrated clearly by the range of projections produced by the Department of Energy for the Sizewell Inquiry (see diagram above). Official energy demand forecasts have been revised consistently downwards. In 1976, for example, projected primary energy demand for the year 2000 was 420–760 million tons of coal equivalent (mtce); by 1982, the range had fallen to 328–461mtce.[2] Such large reductions in projected demand seriously called into question the need to press ahead with ambitious energy supply programmes. Given the obvious room for improvement in energy efficiency, investment in conservation looks an increasingly attractive alternative. A scenario developed by Gerald Leach and his colleagues at the International Institute for Environment and Development suggests that with energy conservation measures already proven or in an advanced stage of development, Britain could achieve economic growth of 3.5 per cent per annum and still require only marginally more energy in the year 2020 than it now consumes.[3] The House of Commons Select Committee on Energy has recently called for more serious attention to be given to the conservation option.[4]

A third basic tenet of energy policy is that objectives should be achieved, as the Green Paper puts it, 'at the lowest practicable cost to the nation'. This seems rational enough – policies should maximise the social benefit/cost ratio. The problem is that we cannot identify such policies unambiguously. Many of the social and environmental costs of alternative options are unidentifiable, unquantifiable and fall unevenly across social groups, space and time. Assessing the costs and benefits of policy options is a task that is fraught with problems.

'Security of supply' has been another enduring principle though it has not always been strictly adhered to. Interesting questions concern the extent to which security is related to *diversity* of supply and to the proportion of imported energy as opposed to indigenous sources in the supply mix. Intuitively, indigenous supplies might seem strategically more secure, but imports in certain circumstances provide greater diversity. Again, these are highly charged political issues. During the 1980s, the Conservative Government justified its determination to press ahead with the nuclear programme, for example, on the grounds of 'diversifying' the source of fuel for electricity supply. In this case dependence on an internal monopoly to provide fuel for 80 per cent of Britain's electricity does not fall within the Government's definition of security.

Finally, the notion of efficiency in energy use has always featured in energy policy rhetoric. In the immediate post-war period, when supply could hardly keep pace with rapidly growing demand, this objective was given some prominence, but throughout the 1950s and 1960s cheap energy made it a secondary consideration. Now it has become a significant issue again. For the Conservative Government, efficiency is something to be brought about largely through rising prices, with minimum interference from the State. Critics advocate a more vigorous and interventionist approach to conservation.

These are some of the factors which have underpinned the formulation of energy policy. Successive governments have emphasised different aspects of the energy economy and have favoured different policy levers. But underlying this diversity has been a remarkable degree of consensus, which has only just begun to be challenged by more radical views of the energy future. Energy policy in Britain has been dominated by the supply industries and their associated trade unions, working closely with Government. Although often in conflict among themselves, these powerful interest groups have effectively excluded others from participation in policy formulation. There are signs that the situation is changing, but the momentum behind the current mode of thinking cannot be ignored; it would certainly be naive to think that policy was based on rational analysis of the prevailing situation and appraisal of all alternative options. But important questions *have* emerged during the 1970s and 1980s concerning the relative emphasis on supply and demand, the lifetime of oil and gas resources, the future of both the coal and nuclear industries and the potential to use renewables. These central energy policy issues are addressed in the remainder of this chapter.

ENERGY DEMAND, 'NEED' AND POLICY OPTIONS

Total primary energy demand in Britain in the mid-1980s was considerably lower than predicted in the preceding decade. It was still, however, a prodigious amount – 328 million tonnes of coal equivalent (mtce) in 1986. Only about one third of energy input actually ends up performing a useful function. Of the remaining two thirds, some is unavoidably lost in conversion and distribution. Much, however, is simply wasted. As a society, we are still extremely profligate with energy resources (photo 100).

Demand for energy is normally broken down by sector and by fuel. In spite of restructuring and de-industrialisation, industry is still a major consumer (28

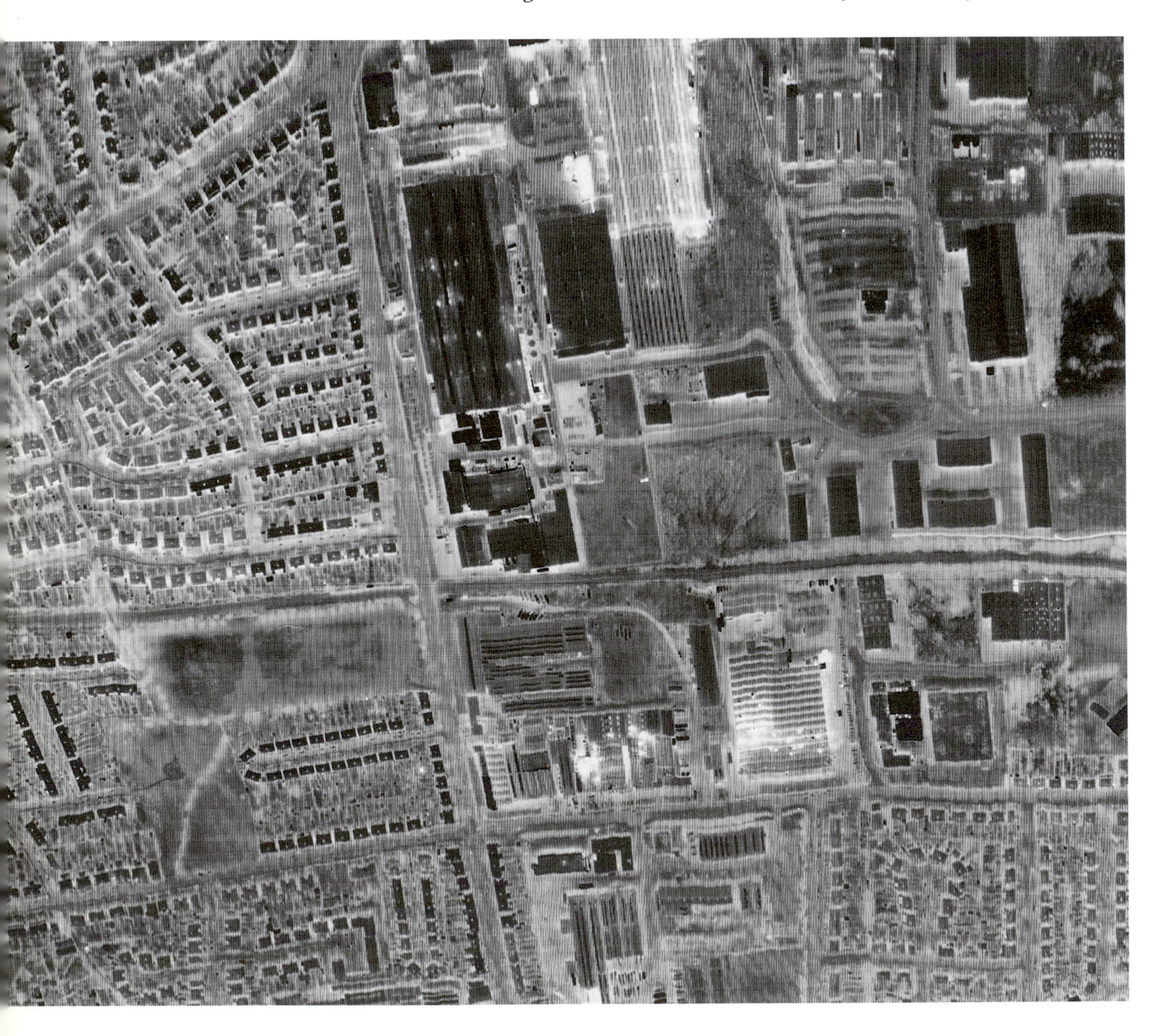

per cent of total final energy consumption), but domestic consumption is now slightly greater (30 per cent).[5] Most final consumers use a mix of energy sources, though gas is increasingly important in the domestic sector (60 per cent of final consumption), and transport is still 99 per cent dependent on oil, a fact with some uncomfortable implications. Final demand can also be broken down by end use, though these data are more difficult to obtain. More than a third of delivered energy is used for low grade purposes such as space heating, and only about 10 per cent is accounted for by essential electricity (purposes for which only electricity is suitable).[6] Such figures have an important bearing on policy choices, particularly in relation to the role of electricity in the future supply mix. One vision of the future, as oil and gas reserves diminish, has been the electrification of society, with increasing reliance on nuclear-generated electricity to meet all energy needs. France is well advanced along such a path. Critics of this strategy have emphasised the 'inelegance' of meeting low-grade heat needs (such as space heating at about 20°C) using electricity generated at 2,000°C, and have argued that supply should be matched to end use as far as possible.

How is demand likely to change? This question is extremely difficult to answer, since macro-economic indicators have become less reliable and demand is influenced by a multitude of factors. Forecasts have been proved wrong so often that the Department of Energy now shies away from producing them: it has published none since 1982. The question is further complicated by the fact that the answer depends upon, *and* is regarded as an important determinant of, energy policy. Projections of demand growth (in total, or for particular fuels) may dictate the 'need' for expansion of supply, but a vigorously implemented energy conservation policy could considerably reduce the rate of growth in demand. The potential for conservation and its role relative to that of energy supply constitute areas for fundamental disagreement in energy policy. Some see the possibilities for improvements in energy efficiency to be so significant that the British economy could continue to grow for several decades with little or no increase in energy consumption; they see a strong case for government intervention to ensure that this potential is realised. Others consider that conservation should be a matter for individual consumers; they reject a major role for 'policy' in this context and generally regard the potential for energy saving as much more limited – perhaps amounting to 10 or 20 per cent of projected unrestrained energy demand. It is, however, widely acknowledged that the market does not always give the correct signals to energy consumers, who for all kinds of reasons tend to under-invest in energy efficiency. And some investments (for example pressurised water reactors (PWRs)) have been subject to less stringent criteria (in terms of expected rates of return) than others, such as combined heat and power generation (CHP).[7]

Even if demand remains relatively stable, non-renewable resources are being depleted and many energy supply facilities, particularly power stations, will reach the end of their useful lives within the next two decades. The one thing that is certain is that Britain will need energy in the future, and even a rigorous conservation policy would not eliminate the necessity for new supply facilities. We now turn, therefore, to some of the important questions about the way in which energy needs in the twenty-first century might be met.

100 *opposite* Aerial thermographic image of part of Cardiff, South Wales. A considerable amount of energy is wasted in the United Kingdom, in spite of the fact that relatively simple energy conservation measures often have very short pay back periods. Heat loss can be graphically demonstrated by infra-red thermography, which is similar to photography except that the picture shows heat instead of light in a black and white image, warm areas appear lighter and heat losses from poorly insulated buildings can be identified. Thermographic surveys can be very useful for 'raising consciousness' about the need for energy efficiency as well as accurately pinpointing sources of heat loss from buildings. This image of part of Cardiff, showing factories interspersed with residential development, has been used in exactly this way following Cardiff's designation as an 'Energy Action City'.

TREASURE FINDER

OIL AND GAS – A WINDFALL FOR HOW LONG?

In two decades since oil and gas were first discovered in the North Sea, the United Kingdom has become the fifth largest oil producer in the world (photo 101). In 1985, oil and gas production accounted for about 5 per cent of gross national product (GNP), employed 29,000 workers offshore (with many more in related industries) and provided the Exchequer with an income of £11.5 billion. There is also a rapidly growing interest in on-shore oil, with well over 50,000 square kilometres under licence for exploration, appraisal or development by the mid-1980s.

The discovery of indigenous hydrocarbon resources on this scale was a remarkable stroke of good fortune for the UK, with clear balance of trade and strategic advantages, and a major potential stimulus to GNP, industry and employment. But whether such benefits are realised depends very much on the economic climate and on policy choices, for there are also risks of overheating the economy, making exports uncompetitive (because of highly valued sterling) and squandering oil revenues. Depletion policies, the fiscal regime within which companies operate, the degree of state involvement and use of revenues have all been contentious topics during Britain's oil era. Whether the North Sea has, on balance, been beneficial for the British economy to date is a matter for continued and often heated debate, though it is almost certainly too soon to reach a definitive conclusion.

The oil and gas deposits of the United Kingdom Continental Shelf (UKCS) (including those onshore) are finite resources. Recent estimates of total recoverable reserves range from 1,962 to 5,392 million tonnes of oil and from 2,406 to 4,506 billion cubic metres of gas.[8] The large range in the figures reflects the uncertainty of reserves estimates (especially for 'undiscovered' resources); geology, technology and prices will interact to determine the quantities of oil and gas eventually discovered and produced. The consensus view, however, is that the most promising areas have now been explored. Although major new finds cannot be ruled out, future discoveries are likely to be in smaller fields and recent figures suggest that UKCS production could well be at maximum in the late-1980s. Comparing reserves estimates with current production suggests a remaining lifetime for UKCS oil and gas measured in decades, but the slope of the depletion curve from now onwards is virtually impossible to predict. Some dissenting voices to the conventional wisdom – most notably those of Peter Odell and Ken Rosing – maintain that oil and gas resources of the European continental shelf have been seriously underestimated, and urge greater incentives for exploration and production to maintain self-sufficiency until well into the next century. Even optimistic estimates suggest, however, that in the medium term we must look to replacement for indigenous oil and gas in the UK energy economy.

The lifetime of a resource and the period of self sufficiency are open to policy manipulation. For the private producer the optimum depletion rate depends on the rate of appreciation in value of the resource in the ground and the rate of return on revenue acquired by extracting it. If the former is the greater (for example, if substantial oil price increases are anticipated), then a slower depletion rate is favoured. If the latter is higher, then it makes sense for the producer to

101 *opposite* Brent oil and gas field, North Sea. The Shell/Esso 'D' production platform is an enormous concrete gravity structure with three legs. The accommodation vessel 'Treasure Finder' can be seen in the foreground and several other production platforms are visible in the background. Extraction of oil and gas from this hostile environment is much more difficult, dangerous and expensive than exploitation on land or in shallow coastal waters. The probability of death per man year for workers on installations in the North Sea is about one in a thousand – some four times the figure for coal miners.

Production costs (1985$) average about $9 per barrel for existing fields and will be about $18 per barrel for fields now under development. Prices are set by the world market, and the steep fall to single figures in the mid-1980s caused considerable consternation and curtailment of exploration activity. A major issue now is the fate of these huge structures at the end of their useful lives. Marine interests would like to see full decommissioning of all platforms, but this could cost oil companies and taxpayers at least £4.5 billion.

extract the resource more rapidly. But the socially optimum depletion rate may be less than the private one because of environmental externalities, attitudes towards future generations or the perceived need for a strategic reserve. If so, there is a case for intervention, though some economists maintain that given the enormous uncertainties involved there is no guarantee that depletion policy would move us any closer to an optimum than 'the market'.[9] Whatever the academic arguments, the UK has never adopted a depletion policy in any true sense. In the initial stages of North Sea development successive governments sought to exploit the resources as rapidly as possible. By the mid-1970s a more conservationist attitude had developed and in 1975 the then Labour Government introduced reserve powers to control the rate of depletion. There was little disagreement between the major political parties on this issue and a general consensus that the reserve powers might become necessary during the 1980s, after achievement of self-sufficiency. In the event, the oil glut of the 1980s ensured that the powers remained largely unused. Depletion has become less of an issue and critics have focused instead on what many regard as misuse of oil revenues to finance high unemployment or to invest abroad following the abolition of exchange controls.

102 *opposite* Steel platform construction, Nigg Bay, East Ross. Platform yards were among the most controversial oil related developments. They require between 50 and 80 hectares of land and a large labour force, involving major physical and social impacts in rural areas. The workforce at the Nigg yard grew from zero to 2,500 (20 per cent of the East Ross labour force) in five months, and emergency accommodation had to be provided on ships which can be seen adjacent to the construction site in the photograph.

Seven sites were eventually established for platform construction. The number of applications was much greater, but since demand for platforms was seriously over-estimated not all were constructed and not all existing facilities are still in use. One yard at Portavadie, Argyle, developed using the compulsory purchase powers of the 1975 Offshore Petroleum Development Act, has never received a single order. The Nigg site, owned by Highland Fabricators, was more successful. Steel platforms can be constructed in shallow water (up to 10m) so sites were in the east, close to offshore fields. Concrete platforms required the deeper water of the West Coast, where applications involved significant conflict with amenity.

The relationship between governments and companies over control of oil and gas resources has been accurately described as one of 'antagonistic interdependence',[10] in which clearly divergent interests are tempered by mutual recognition of the need for a working relationship and cooperation. Control is complex, but (apart from the kind of direct policy intervention discussed above) basically falls into three categories – licencing for exploration and production, taking of revenue through royalties and taxation (including petroleum revenue tax, currently set at 75 per cent of profits), and state ownership or participation. Licencing and taxation regimes have changed little between governments of different political persuasion, though in line with the less interventionist stance of the Conservative Government some blocks are now 'auctioned' (awarded to the highest cash bidder) in licencing rounds instead of being allocated by civil servants after scrutiny of companies' exploration programmes. Ideological differences have been most in evidence in attitudes to state ownership and participation. The British National Oil Corporation (BNOC), established in 1975 by a Labour Government and given (from 1978) an equity share greater than 51 per cent in all new developments, was partly denationalised by the Conservatives in 1982 and finally dissolved in March 1986. The British Gas Corporation lost its monopoly purchasing powers in 1982 and has now been privatised.

As well as having major implications for the economy as a whole, development of UKCS resources has had important socio-economic and environmental impacts regionally and locally. During the 1970s the main focus of activity was in Scotland when there was an intensive phase of development associated with exploration, initial extraction and transport, requiring drilling rigs, production platforms, pipelines, terminals and a whole range of port, service and administration facilities (photos 102 and 103). Developments were welcomed in many areas because of the employment and prosperity they were expected to bring. This was especially true in the depressed lowland parts of Scotland, but in more remote areas reaction was sometimes ambivalent. Jobs and prosperity would be welcomed, but not if

they undermined traditional industries then left raised expectations and economic depression when the 'boom' was over. Unfortunately, the distribution of oil-related jobs did not favour the areas in greatest need. By 1978, oil jobs represented only 1 per cent of total employment in Strathclyde, whereas in Grampian and Highland Regions they represented 7 per cent and 9 per cent respectively. Yet these rural areas had well established traditional industries and unemployment rates below the Scottish average. There was understandable concern that oil activity would cause a decline in local manufacturing employment, which proved well founded in some places.[11] Infrastructure was a major problem in remote communities and 'emergency' provisions had often to be made to house construction

103 On shore terminal, St Fergus, Aberdeenshire. The St Fergus complex shared by Total Oil, Shell/Esso and the British Gas Corporation (BGC) under construction in 1975; the sharing of sites by different operators avoids proliferation of terminals along the coast (see inset). The pipeline landfall can be seen clearly in the foreground. Natural gas (methane) from the Brent and other fields is separated here from natural gas liquids (NGL), stabilised, pressurised and transferred to BGC for injection into the grid. The plant at St Fergus is highly automated, and although the site occupies over 200 hectares of land and produces on average 11 million cubic metres of gas and 5,850 tonnes of NGL per day, it is run by a permanent staff of about 100 people.

In spite of its scale, St Fergus was not a controversial development, but the smooth and rapid progress through formal planning procedures (in 1973) was associated with a great deal of 'behind the scenes' activity. BGC's preferred, but much more contentious, site on the Loch of Strathbeg was abandoned in favour of St Fergus, considerable effort was made to minimise impacts of construction and operation and liaison committees were established to deal with environmental and amenity matters.

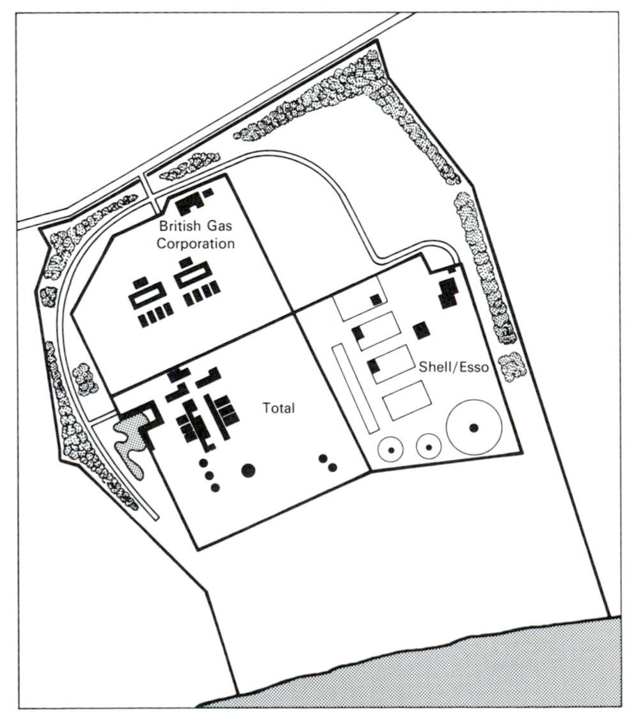

Plan of the St Fergus terminal. (Source: based on Shell UK, *St Fergus Gas Plant*.)

workers. Socially and economically therefore, the oil boom has been a mixed blessing for Scotland and has certainly not resulted in a lasting transformation of the Scottish economy. Furthermore, places like Aberdeen which reaped benefits in the 1970s are now feeling the cold winds of change most acutely.

The environmental impact of North Sea-related developments was potentially alarming, and in the haste to bring oil and gas ashore some damage to fragile areas was inevitable. Legislation (the Offshore Petroleum Development Act 1975) was introduced to curtail planning and consultation procedures in case these proved to be too time consuming, though in practice it remained largely unused for this purpose. Around one hundred applications for oil and gas related developments were submitted in Scotland during a relatively short time span in the mid-1970s. Planning authorities, especially in remote, sparsely populated areas of high amenity value, faced severe conflict between demand to process planning applications rapidly and to provide sufficient sites for development, and the need to assess proposals properly and fit them into some overall strategic framework. The latter proved difficult to achieve, in spite of the issue of several sets of guidelines by the Scottish Office. Some local authorities – notably the old Zetland and Orkney County Councils – adopted innovatory approaches to planning pressures and took legislative and other steps to ensure that the community participated in decisions and benefited to the maximum extent from oil-related developments, but such experience was the exception rather than the rule.

Unless there are significant discoveries in 'new' areas of the UKCS it is unlikely that much further infrastructure on land will be needed. On-shore oil deposits, however, are likely to generate controversial development pressures both for exploration drilling and for production. Already there have been a number of skirmishes – most notably over Shell's proposals for exploration drilling in the New Forest in the early 1980s, though there have also been relatively successful developments as at Wytch Farm in Dorset (photo 104).

To sum up, Britain is likely to be producing oil and gas for several more decades and could be self-sufficient into the twenty-first century. But within the foreseeable future these resources will have to be replaced, and before that it is likely that they will be reserved for uses in which they are least easily substituted (e.g. transport and petrochemicals). Whether the benefits of exploiting the UKCS outweigh the costs – both locally and nationally – is a question which must remain to be answered in the future. Meanwhile, it is important to consider questions about the resources which will be available as oil and gas are depleted. One such issue, which has recently dominated the energy debate, is whether new electricity generating capacity should be built for coal, or nuclear power, or both.

COAL AND NUCLEAR POWER

Coal and nuclear power need not occupy the same niche in the energy market, but it is widely perceived that the futures of these two energy sources are inextricably bound together. Economics, environment, safety and employment are all important issues in the debate, underlying which are two fundamental questions – should Britain maintain a civil nuclear power programme at all, and can British coal be produced at prices that are competitive with its substitutes, including imported coal?

104 Wytch Farm, Purbeck Peninsula, Dorset. The costs of finding and producing onshore hydrocarbons are very much lower than those for resources located offshore. The onshore oil industry in the UK experienced something of a 'boom' as world oil prices fell in the 1980s. The photograph shows the most significant onshore oil field developed jointly by the British Gas Corporation (BGC) and British Petroleum (BP), now operated by BP. The field gathering stations and four well sites can be seen (see accompanying drawing). The gathering station receives oil from the production wells, separates it from gas and water and stores it pending delivery to the rail terminal (not shown). 'Nodding donkeys' (the pumps) can be seen at the nearest well site.

The Isle of Purbeck is an area of great beauty and ecological significance. Development was carried out in full consultation with the local authority and other interested parties and is widely acknowledged to be a model of environmental sensitivity. Nevertheless, the increasing scale of onshore activity and the granting of licences to operators with no previous experience gives rise to concern about potential conflicts with conservation and amenity.

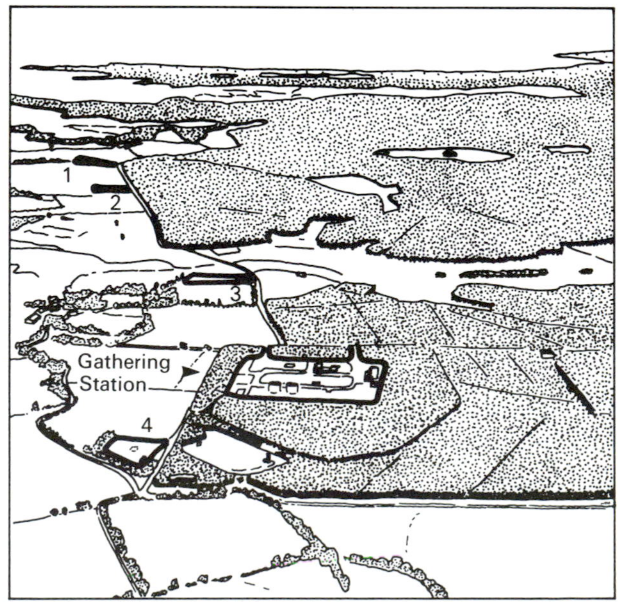

Plan of site in photograph 104 with well sites shown by numbers.

105 Sizewell, Suffolk Coast. The Sizewell 'A' nuclear power station is one of nine 'Magnox' reactors, using uranium metal fuel clad in magnesium alloy, graphite moderator and carbon dioxide coolant. Coastal (or lakeside) sites are essential for access to cooling water. These reactors have no secondary containment vessels.

The application for Sizewell 'A' in 1959 was approved within a year, with few objections and no inquiry. The station was commissioned in 1966. In contrast, consent for the Sizewell 'B' pressurised water reactor (PWR), the site for which is highlighted on the photograph, took well over five years to obtain and involved a hugely expensive 340-day public inquiry.

The Magnox programme was generally regarded as successful, if expensive (unlike the second, Advanced Gas Cooled Reactor (AGR) programme, which ran into considerable time and cost overrun problems). But the Magnox reactors will come to the end of their useful lives in the 1990s, and decommissioning costs would have been an unacceptable liability for the private sector. All nuclear power stations have been retained in public ownership.

106 *opposite* Selby Coalfield, Yorkshire. Coal was discovered under this predominantly agricultural part of Yorkshire in 1972, and in 1974 the NCB sought planning permission to mine the rich Barnsley seam at the rate of 10 million tonnes per annum – the first major application within the framework of 'Plan for Coal'.

In 1988 the complex is well on its way to completion. Coal is brought to the surface via a pair of drift mines at Gascoigne wood (shown in photograph). There is also a shaft at Cawood and five satellite shafts to give access to miners and materials. Coal is removed by rail (trucks visible in photograph).

Local consternation at development on such a scale was great enough to cause the Secretary of State for the Environment to 'call in' the application. The main issues at the public inquiry were environment, amenity and infrastructure. Interestingly, in the aftermath of the oil crisis, the national 'need' for the coal was not in dispute. Consent was granted subject to stringent conditions including a requirement for coal 'pillars' to be left under the town of Selby and a maximum level of subsidence of 0.99m throughout the area.

The cheap Middle Eastern oil which flooded the world market in the 1950s and 1960s affected both the ageing coal industry and the embryonic civil nuclear power industry in Britain. After its post-war peak in production of 224mt in 1955, the coal industry declined rapidly, in spite of protectionist measures including a coal import ban (1959–69) and an import duty on heavy fuel oil (see diagram on p. 190). The nuclear industry failed to expand as expected and by the 1970s had run into difficulties with construction and commissioning of advanced gas cooled reactors (AGRs), as well as the problem of growing public hostility (photo 105). But the four-fold increase in the price of oil in 1973/74 seemed at first to transform the prospects for both industries and ambitious expansionary programmes were hastily assembled. The 1974 'Plan for Coal', agreed by the National Coal Board (NCB), the National Union of Mineworkers (NUM) and the then Labour Government, aimed first to stabilise production, then to raise it from 130mtpa in 1973 to between 135 and 150mtpa by 1985. Progress was reviewed in 1977 and even more ambitious plans to raise output to 170mtpa by the year 2000 were agreed.[12] These plans, broadly endorsed by the 1978 Energy Policy Green Paper, implied profound changes to the geography of coal production in Britain. They envisaged the closure of old capacity at an average rate of 2mtpa, massive new investment in certain existing mines and the opening of new capacity (between 1985 and 2000) at a rate of 4mtpa – the equivalent of two new deep mines every year. Closures would be concentrated in the older, peripheral coalfields, new investment in the central coalfields and much new capacity in areas hitherto untouched by mining activity (photo 106). The social and environmental implications were of great significance.

Plans for the nuclear industry were no less ambitious. In evidence to the Royal Commission on Environmental Pollution (RCEP) in 1976 the United Kingdom Atomic Energy Authority (UKAEA) envisaged a twenty-fold increase in nuclear generating capacity by the end of the century and a further four-fold increase, mainly from fast reactors, in the following thirty years![13] The ordering of one or two new nuclear power stations per annum was considered the 'norm' and in 1979 the Secretary of State for Energy announced a third programme of ten stations, the first of which was to be a pressurised water reactor (PWR).

Ten years later, things looked very different. Plan for Coal had been abandoned, its ambitious production targets replaced by a new market-orientated strategy, the third nuclear power programme had scarcely begun, and coal and nuclear power, far from expanding in tandem to meet growing energy demand, had been drawn into antagonistic competition. What happened to bring about such a complete change of circumstances? Major factors were stagnating energy demand, a much easier world energy supply situation, and the resulting strong downward pressure on oil (and internationally traded coal) prices from the early 1980s onwards. Industrial markets for coal proved very resistant to recapture, the steel industry suffered deep recession and the need for new coal technologies (such as coal gasification) receded into the next century. The failure of alternative markets to materialise made the electricity generating sector, which takes around 70 per cent of British coal production, central to the future of the industry (photo 107). But electricity demand growth was slower than anticipated in the 1970s, the policy of the CEGB (fully supported by the Government), was to 'diversify'

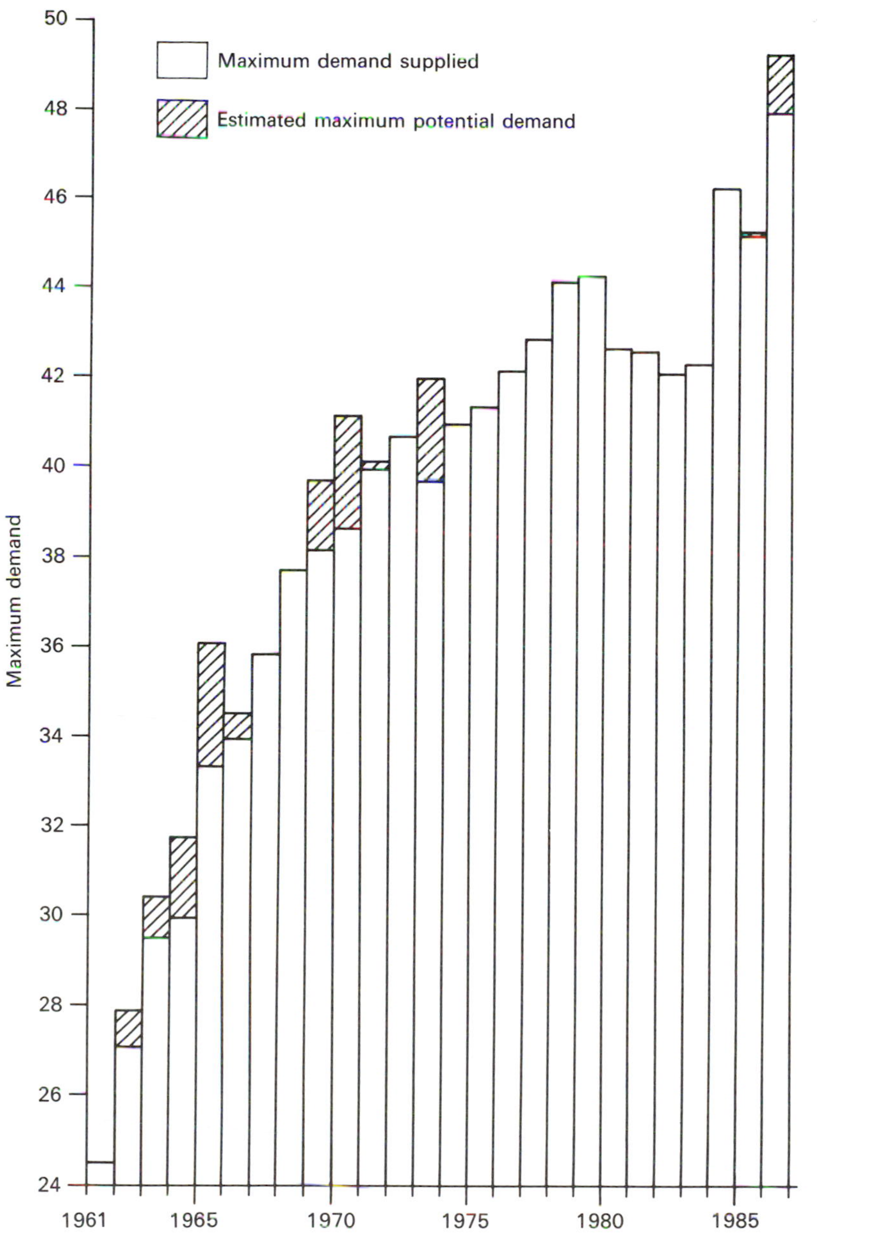

Electricity demand 1961–87. (Source: Electricity Council.)

107 *opposite* River Trent power stations, Nottinghamshire. Cottam power station (2,000mw) is in the foreground, with West Burton ('A' and 'B') behind (the latter is potentially the site for a new coal-fired station). Many power stations were constructed in the 1960s to meet rapidly growing electricity demand. Coal is delivered directly from the mines by 'merry-go-round' trains. The Trent provides cooling water.

The high stacks which can be seen at each power station became standard after the 1956 Clean Air Act, to minimise ground level concentrations of sulphur dioxide and other emissions. It has taken several decades to realise that this measure shifts a pollution problem in time and space, contributing to 'acid rain' in other European countries. New coal-fired power stations – and a number of existing ones – must now be equipped with flue gas desulphurisation (FGD) equipment to reduce acid emissions at source.

The inland location of many power stations on the coalfields affords the British coal industry some protection against cheaper imports, but steeply falling fossil fuel prices enabled the CEGB to negotiate a 'new joint understanding' with British Coal (BC), to run to 1991, under which the CEGB obtains increasing amounts of coal from BC at prices reflecting those of imported oil and coal.

capacity (currently 80 per cent coal) to achieve a supply mix of 50 per cent coal, 42 per cent nuclear and 8 per cent renewables by the year 2020, and the nuclear lobby insisted that a viable nuclear industry must have a steady stream of orders for new power stations. Conflict was inevitable in these circumstances. Issues of relative economics, environment and safety were all drawn into the debate and many of the arguments were rehearsed at the 340-day inquiry into the CEGB's application to construct Britain's first PWR at Sizewell on the Suffolk Coast (photo 105).

The economics of electricity generation is a difficult and contentious subject, and in the recent past both the Monopolies and Mergers Commission and the House of Commons Select Committee on Energy have been critical of the CEGB's accounting procedures. At the Sizewell inquiry objectors vigorously contested the assumptions behind the CEGB's cost estimates, but while the inspector, Sir Frank Layfield, agreed that the CEGB figures were optimistic, he concluded that there was only a one in forty chance that a coal-fired power station would generate electricity more cheaply than Sizewell 'B' (and a one in four chance that an AGR would do so).[14] The sensitivity of such conclusions to variables which are inherently difficult to predict was starkly illustrated when the Secretary of State for Energy issued his decision letter, in which because of steeply falling fossil fuel prices the probability of coal being a 'better buy' had been drastically revised to one in seven. The Government's proposals to privatise the electricity supply industry included a *requirement* for the Area Boards to purchase a specified proportion of nuclear (strictly 'non-fossil') generated electricity – hardly a necessary step if this option were clearly the most economic. During 1989 – when the nuclear industry was exposed to the City's economic appraisal in the run-up to privatisation – first the ageing Magnox reactors and then the AGRs and the Sizewell B PWR were removed from the flotation. The idea that nuclear power produces cheap electricity became impossible to sustain.

108 *opposite* Opencast coal mine, Hauxley, Northumberland. Opencast mining is amongst the most profitable and the most environmentally destructive of British Coal's activities. The photograph gives some idea of the scale and impact of a typical site. To maintain current output (about 14mtpa), the Opencast Executive (OE) must acquire 2,000–4,000 hectares of land annually and open a new site, on average, every three weeks. Production is capital intensive (note the large drag line, centre right) and is contracted out to the civil engineering industry. Although controls and restoration techniques have greatly improved, the impact on local communities in terms of noise, dirt, traffic and social upheaval can be very considerable. Opportunities to mine and reclaim derelict land (e.g., old spoil heaps) are diminishing and the OE increasingly seeks green field sites; close encroachment on agricultural land can be seen in the photograph. Environmental groups argue that production should be curtailed except in special circumstances (a view endorsed by the Commission on Energy and the Environment 1981). The OE recently announced its intentions to expand production to 18mtpa or more.

The future price of coal, as well as that of nuclear generated electricity, has been a major uncertainty. This question has ramifications going well beyond the economics of electricity generation, but which must be considered here since they are crucial for the future of the British coal industry. Deep-mined coal production in the UK is expensive in relation to that in other established producing countries, such as the USA, South Africa and Australia. Other producers, including China and Colombia, are likely to be contributing cheap coal to the world market in the near future, and there may be reductions in bulk sea transport costs. The British industry (like its counterparts elsewhere in Europe) has traditionally enjoyed subsidy and protection from imports on strategic and social grounds. It is also afforded some 'natural' protection from cheap imports by the lack of deep water port facilities and the inland location of many of its major customers. Profound changes in attitude to the coal industry have been apparent since the 1979 election of a Conservative Government, drawing few votes from mining constituencies and traditionally less sympathetic to 'lame ducks' and to less prosperous regions. The transition to a new and stringent financial regime has been marked by bitter conflict, culminating in the year-long mining dispute of 1984/5, the outcome of which permitted the Government and British Coal (BC – renamed in 1986) to press ahead with 'streamlining' of the industry. The declared aim is to make it efficient and competitive – with other fuels and quite possibly with imports – and, many believe, to prepare it for privatisation. Any colliery producing coal at an operating cost of more than £1.65 (1985) per gigajoule (GJ) is now put into the review procedure for closure, and costs of no more than £1.50 per GJ were the stated aim of BC's 'New Strategy for Coal'.[15] In fact that target was soon exceeded, with costs reported as £1.43 per GJ at the end of the financial year 1985/6.

Both pit closures and the extent of imports are subjects on which strong and polarised views are held (see further reading for a more detailed exposition). At one extreme are those who would prune the industry fairly ruthlessly and expose it to competition to force it into solvency and efficiency. Proponents of this view have largely held sway since the end of the strike: the workforce was reduced by 45 per cent between 1983 and 1987 and productivity increased by nearly 70 per cent.[16] At the other extreme are those who see coal as a vital national resource rather than a commodity for the free market, and who urge that policy be based on a wider 'social cost accounting' procedure. If it is assumed, for example, that redundant miners will not be re-employed, so that 'rationalisation' incurs measurable costs in terms of redundancy pay, unemployment benefits, social security, regional policies, and tax income (as well as production) foregone, the economic rationale for closing even grossly 'uneconomic' pits, at least in the short term, diminishes. The unmeasurable social costs of unemployment and loss

of communities, and the environmental costs of opening new capacity in green field sites should also be included in the calculus. Arguments like these have led some critics of current policy to propose that deep mined production be protected by henceforth ordering only coal-fired power stations, banning imports of coal (and electricity) and curtailing the profitable but environmentally destructive opencast sector which currently produces about 15mtpa and is set, very controversially, for expansion (photo 108). Such measures, it is argued, should be supplemented by positive incentives for coal use in the industrial sector.[17] While the argument raged, between February 1984 and November 1986 54 pits were closed. Although individuals have received generous redundancy settlements, progress towards solving the wider social problems of the older coalfields has been meagre. Productivity, on the other hand, has reached new records and BC's financial performance has improved markedly. Commenting upon this 'remarkable and commendable progress' the Select Committee on Energy noted the 'substantial' short-term public costs of the industry's transformation and the 'heavy burden' which it was placing upon coalmining communities. They also concluded that structural change towards higher productivity and lower costs was far from complete.[18]

Economic considerations are one factor in the elusive cost-benefit analysis of energy options. The other most important issue for debate has been the social and environmental risk associated with different fuel cycles. It is a debate characterised by considerable scientific uncertainty in which each new piece of evidence tends to be used as ammunition to support or refute the views of the various

109 Merthyr Tydfil, South Wales. The legacy of dereliction from past mining activities still has a powerful impact on the image of the coal industry. Poor environmental quality is a deterrent to much-needed new investment in communities suffering from the restructuring of the industry. Efforts are being made to improve the landscape, but the resources available are scarcely adequate given the scale of this problem in Britain's mining areas. Merthyr Tydfil in South Wales is typical. Derelict opencast workings can be seen towards the top right of the photograph; the southern part of the site has been restored after opencasting in the period 1975–87. Between the opencast workings and the road are colliery disposal tips, now undergoing progressive restoration. The relics of other heavy industries also blight the coalfields. The great Dowlais Ironworks has certainly left its mark on Merthyr: remains of ironstone workings (centre, top), the site of the works itself (next to road, opposite colliery tips), the Dowlais Great Tip (once a massive slag heap from the ironworks) and red ash shale tips (bottom left) are visible. The first two have been reclaimed, and the last two are programmed for restoration by the mid-1990s.

110 Belvoir Castle and the Vale of Belvoir, Leicestershire. Belvoir was the second major 'green field' site application to come forward under Plan for Coal. In 1978 the NCB submitted planning applications for deep mines, pithead facilities and spoil tips at Hose, Asfordby and Saltby. The application was 'called in' early in 1979 and a lengthy public inquiry ensued. Belvoir encapsulated many of the classic conflicts involved in energy developments. The need for coal had emerged as a contentious issue by 1979 and intensified the debate about the environmental and social impact of the development in an 'unspoilt' rural area. Ranged against the NCB were local authorities, the Countryside Commission, the Council for the Protection of Rural England and the 'Alliance', representing farmers, local residents and parish councils. The Duke of Rutland, whose ancestral seat is Belvoir Castle, was prominent in the anti-NCB campaign. Consent for mining at Hose was refused by the Secretary of State for the Environment in March 1982, but consents for the pits at Asfordby and Saltby have subsequently been granted and construction is in progress.

protagonists. Risks cannot always even be identified, let alone quantified to permit any rational comparison – though this has not prevented people from trying. In any case, in a democracy, it is public *perception* of risks rather than assessment by 'experts' which often determines what is politically acceptable.[19]

No source of energy is free from risk or environmental impact. Coal has a legacy of accidents, disease and dereliction to live down (photos 109 and 110), requires green field sites in rural areas for new developments, and is associated with serious transfrontier and global pollution problems like acid rain and the greenhouse effect. Problems (with the possible exception of carbon dioxide accumulation) *can* largely be overcome but the costs are high and there is conflict over who should bear them. The nuclear fuel cycle is claimed by its supporters to be clean and safe by comparison. But its advocates have been unable to overcome two major anxieties – fear of radiation, released during normal operation or by accident, and fear of a nuclear disaster of catastrophic proportions. Three Mile Island, pollution from the Sellafield nuclear reprocessing plant in Cumbria (photo 111), unexplained cancer clusters near nuclear installations and, most tragically, the Chernobyl nuclear accident of April 1986 have all belied repeated reassurances from the nuclear industry and governments; many people have greater faith in Murphy's Law.[20] In addition there is concern about the civil liberties implications of the 'plutonium economy' associated with a fuel cycle involving both conventional and 'fast' reactors (photo 112),[21] and about the still unresolved problems of decommissioning and radioactive waste management. Waste is currently stored at power stations, at Sellafield and at a nearby shallow burial site (Drigg). A test drilling programme to investigate sites for high-level

waste disposal was abandoned in face of vehement opposition in 1981. In the wake of the Chernobyl disaster it was first announced that four sites under investigation for intermediate level waste disposal would take only low-level waste, then, in the run-up to the General Election in June 1987, the programme was cancelled altogether. It is now proposed that low- and intermediate-level wastes be buried with high-level waste and further investigation is being undertaken of the suitability of sites at Sellafield and Dounreay, which effectively emerged as the least politically unacceptable locations for a disposal/storage facility.

It is clear is that the prospects for coal and nuclear power in Britain cannot be evaluated in isolation from each other or from the wider social context in which energy decisions are made. Pressures to rationalise the coal industry, or to phase out nuclear power have been strongly resisted in a climate of de-industrialisation and regional decline (see chapter 5). This conflict is familiar to observers of the coal industry, but it came into sharp focus for the nuclear industry too in September 1986 when the Labour Party at its annual conference adopted a policy of phasing out nuclear power. Tens of thousands of people are employed by the nuclear industry in Britain, nearly 30,000 by the UK Atomic Energy Authority and British Nuclear Fuels Ltd, plus relevant staff of other organisations like the CEGB. While the risks and environmental impacts associated with both the coal and nuclear fuel cycles underline the attractions of energy conservation and should encourage more serious consideration of renewable resources, the political realities of entrenched interests and immediate economic and employment implications in the regions will continue to be important factors in policy decisions. The Government clearly intends, however, that decisions about energy supply should in future be less a matter for policy than for the market.

The privatisation of the electricity supply industry will certainly have an effect on fuel choice, though the precise implications are as yet unclear. But it is likely that both coal and nuclear power will find themselves competing with gas in electricity generation, the latter having significant advantages in environmental terms, and that investment in energy efficiency will seem an increasingly attractive alternative to large-scale and long-term commitment of capital to major supply facilities.

111 *opposite* Sellafield nuclear reprocessing plant, Cumbria. Sellafield (formerly Windscale) is Britain's most controversial nuclear facility. Reprocessing of spent fuel involves the separation of useful uranium and plutonium from waste products. More than 25,000 tonnes of Magnox fuel have been reprocessed at Sellafield and a new thermal oxide reprocessing plant (THORP), to deal with AGR fuel, is under construction.

Sellafield became the focus of attention in 1983 when a television documentary suggested a link with abnormally high childhood cancer rates in the vicinity of the plant. Soon afterwards, beaches were closed to the public after contamination by an accidental release of radioactivity. In the political storm that followed, permitted discharges to the Irish Sea were reduced to one tenth of their previous level.

The necessity for reprocessing is much disputed. Opponents claim that it is not necessary to 'close' the civil nuclear fuel cycle, and they link reprocessing to military proliferation. Proponents argue that the technology is both necessary and highly profitable, providing a valuable source of foreign revenue for Britain.

WHAT PROSPECTS FOR RENEWABLE ENERGY?

Official forecasts typically envisage a minor role for renewable energy sources in Britain during the first part of the twenty-first century, though many environmentalists argue that these technologies could become much more significant if there was a genuine political will to develop and exploit them. A research and development (R&D) programme was established after the first oil crisis, but funding has been on a very modest, if not parsimonious, scale when compared with expenditure on more conventional energy resources: in 1986, for example, only 4.4 per cent of government R&D was on renewable sources.[22] The Department of Energy defends the relatively low level of expenditure on the grounds that development has not reached a stage where large injections of money would necessarily increase the rate of progress; others consider this to be a peculiarly circular argument.

112 Experimental fast reactor, Dounreay, Caithness. Britain's experimental fast reactor is remotely situated on the North Scottish Coast and is a major local employer. Fast reactors convert non-fissile material (such as Uranium 238) into plutonium which can be used as reactor fuel – hence the term 'breeder' reactor. The long-term future of civil nuclear power would involve fast reactors to extend the lifetime of uranium reserves by a factor of about 60. But the fast reactor programme is highly controversial; environmentalists have long expressed disquiet about the civil liberties implications of the 'plutonium economy' and the Treasury has been increasingly concerned about the costs of the development programme. As uranium reserves estimates have risen, the prospects for a commercial scale fast reactor in Britain have receded and in 1989, the Government announced major cuts in the programme's funding. However, Britain participates in a European programme, and the Atomic Energy Authority was recently given consent to construct a demonstration reprocessing plant at Dounreay. A local public inquiry into this proposal in 1985/6 was boycotted by many environmental groups because of its restricted terms of reference.

Assessment of the potential of the renewable resources – biofuels and solar, wind, wave, hydro, tidal and geothermal power (the latter not strictly 'renewable', but usually included) – has been carried out by the Department of Energy's Energy Technology Support Unit. ETSU concludes that hydropower, (photo 113), some biofuels and passive solar energy (building design to take maximum advantage of solar gain) are already economically attractive; on-shore wind (photo 114) tidal and geothermal energy from hot dry rocks (HDR) are very promising, but require more research and development; some forms (for example, wave power or active solar energy) must await a major breakthrough or significant changes in the energy economy to become viable in the UK.[23]

If fossil fuel prices are assumed to rise more steeply, the renewables become more attractive, though in the event of abrupt price 'shocks', those with long lead times (like tidal power) would not be as easy to bring onstream as, for example, energy from biofuels. Another plausible scenario which would favour the renewables (especially those producing electricity) would be a future with restrictions on nuclear expansion and tight controls on emission from fossil fuel plants. It should be noted that ETSU's analysis implicitly accepts the conventional wisdom on nuclear economics, so that the economic benchmark for the electricity producing renewables is 'cheap' nuclear power. If alternative views of nuclear economics are accepted, this again improves the prospects for renewable technologies.

It is clear that the technical potential of the renewables is enormous – up to 270mtce/year (cf. the total 1986 UK energy consumption of 328mtce). But

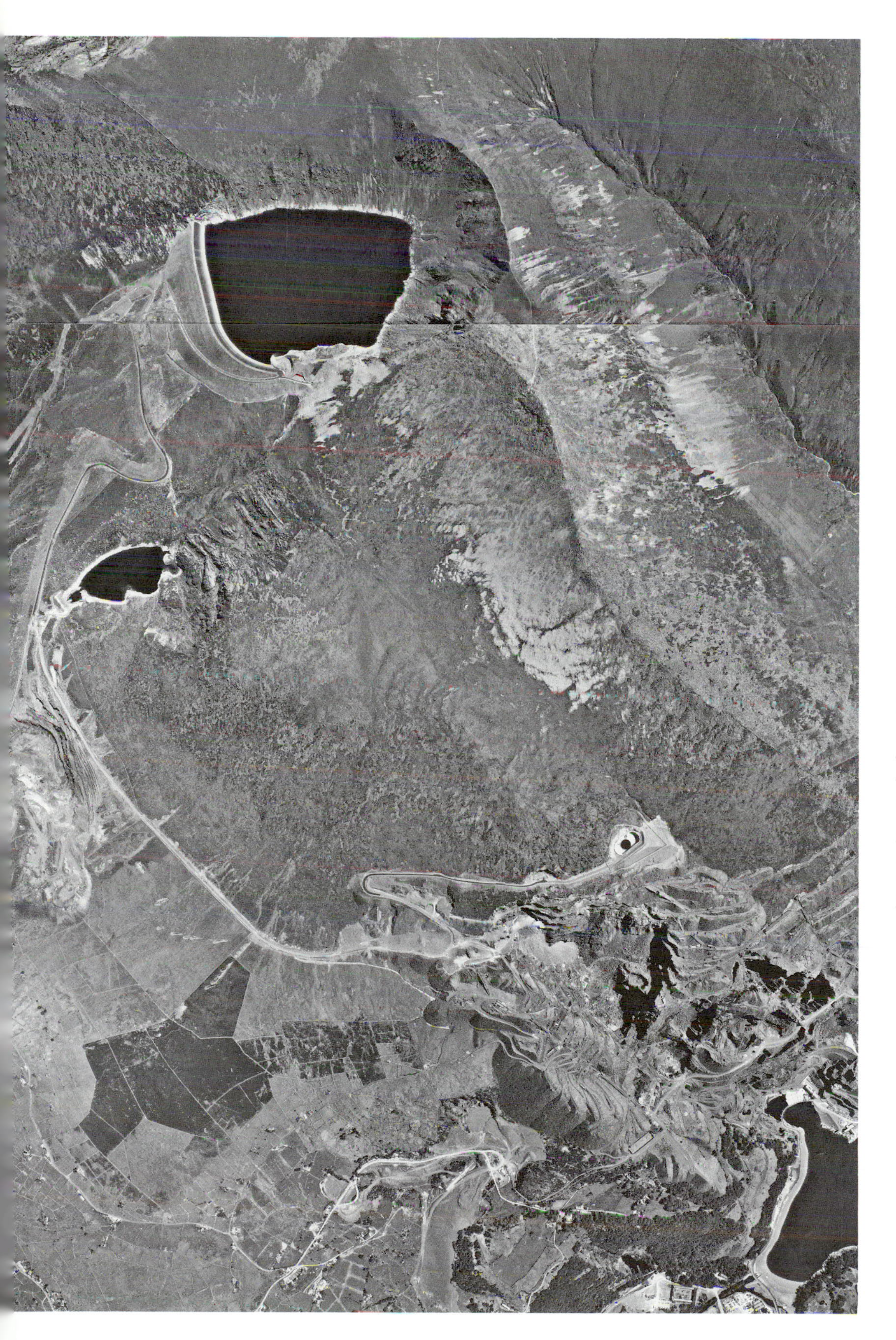

113 Pump storage scheme, Dinorwic, North Wales. The recently completed, £425 million power station at Dinorwic in the Snowdonia National Park is not a conventional hydro-electric station but a pumped storage scheme, designed to help the CEGB cope with diurnal variations in electricity demand. At times of low demand, cheap 'base load' electricity is used to pump water from the Peris Reservoir (bottom right) to the Marchlyn Reservoir over 500m above (top). When demand is high, water is returned through the turbines, which are invisible because the mountain has been hollowed out to house them in an operation involving the excavation of 1.25 million cubic metres of rock. Maximum power can be achieved within a few seconds, and the process is 70–80 per cent efficient. Environmental impacts have been much reduced by 'hiding' the main works, but nevertheless include fluctuating water levels, extensive movement and disposal of materials and visual intrusion. Dinorwic is undeniably an impressive feat of engineering and has become a showpiece for the CEGB. Proposals in the 1970s for similar schemes in the Peak District and in Scotland met strong resistance, and were shelved because of relatively slow growth in electricity demand.

ETSU's estimate of the likely contribution in the year 2025 (in their 'stable' scenario) amounts to little more than 10 per cent of this potential – slightly greater if the Severn barrage (which could provide 5 per cent of the electricity requirements in England and Wales) is built. Much of the difference between theoretical and practical potential can be attributed to the analysis of economic viability. But cost effectiveness is not a sufficient condition for the adoption of a technology (or even a necessary one, if the history of nuclear power is a precedent)[24] – there are also major social, institutional and environmental obstacles to be overcome. One of the most important is the lack of an effective 'lobby'; the renewables do not (yet) include powerful vested interests among their friends. The CEGB is involved with electricity-producing renewables, but is usually anxious to point out that (with the exception of hydro-power), 'at present none of the technologies . . . provide an economic alternative to nuclear power'.[25] One certain exception (even accepting the CEGB's figures for nuclear costs) would be passive solar energy, where significant savings in space heating costs can be achieved with no significant economic or aesthetic penalties. But institutional inertia means that there are no provisions in planning or building regulations to ensure 'energy-conscious' design and most new development proceeds with scant attention to simple principles for harnessing an effectively 'free' resource.

In other cases potential environmental impacts deserve serious consideration. Concern has already been expressed about the inevitable impact of tidal barrages on estuarine ecology and pollution dispersal. The aesthetic effects of on-shore wind generators, many small hydro-schemes or cooling towers associated with HDR resources (many of which lie under areas of high amenity value) would certainly generate controversy over their siting. The renewables lack the potentially catastrophic and irreversible global impacts of fossil fuels or nuclear power (except perhaps for major dam failure), but as implementation of some of the technologies approaches, it becomes apparent that 'renewable' is not necessarily synonymous with 'environmentally benign'.

114 *opposite* Wind generators, Carmarthen Bay, South Wales. Wind power is one of the most promising renewable energy sources. The photograph shows the CEGB Carmarthen Bay power station with the Department of Energy 130kw vertical axis wind turbine in the foreground (supplied by VAWT Ltd), the Balfour Beatty Power Construction Ltd 10kw vertical axis machine in the centre and to the rear the tower and nacelle of the CEGB 300kw wind turbine (supplied by James Howden Ltd). Early in 1988, the government announced plans to spend £28 million on an experimental 'wind farm'.

The concern about energy supply which dominated the 1970s has abated. Fossil fuel prices are low, supplies apparently plentiful and, in spite of the shock of Chernobyl, complacency has again set in. Yet the issue lurks uncomfortably near to the surface. Even if we are fortunate enough to avoid another energy shock, (oil prices turned sharply upwards in response to political instability in the Gulf in 1987), costs (in the widest sense) of meeting energy requirements are going to rise. Indeed, it is the environmental costs of energy production and consumption which now ensure a prominent place for energy on the political agenda. Balancing energy 'needs' against these costs will become an increasingly contentious issue.

One important lesson learnt since the early 1970s is that rigid forecasts of energy demand and blueprint planning for supply are doomed to failure. But to move to the other extreme with those who claim that Britain's energy future should be 'left to the market' is equally absurd, and for Government to have no well-defined policy in this crucial field is an abrogation of responsibility. The motley collection of policies and programmes outlined in this chapter will, if pursued in the short to medium term, lead to continued depletion of UKCS oil and gas, increased use of gas for electricity generation, further restructuring of the coal industry (with some new investment), possible construction of a Severn Barrage by the private sector and tentative developments of other renewables and CHP. And the energy sector will continue to be privatised. All developments are likely to be contentious on social and/or environmental grounds, with intense conflict almost certain to arise over nuclear waste disposal, opencast mining and major estuarine facilities. If demand grows slowly, as seems probable, this will be due more to poor economic performance and restructuring of the economy than to positive demand management.

Whether these policies constitute the most appropriate package has never

Landscapes under threat:
115 *above* Widdington, Northumberland.
116 *opposite, top* Oxfordshire countryside near Witney.
117 *opposite, bottom* Severn Estuary.
None of Britain's energy supply options is free from implications for landscape and the environment. New nuclear power stations may require attractive coastal sites such as that at Druridge Bay, Northumberland, 'shortlisted' by the CEGB in the 1980s, though it is unlikely that any new nuclear power stations will be constructed for some time, if at all. Coal reserves are likely to lie under rural areas with no previous experience of mining: parts of Oxfordshire, for example, were included in the NCBs exploration programme during the 1970s. The 'renewables', too, may have a major impact. Proposals to construct tidal barrages across the Severn and other estuaries have generated considerable alarm amongst nature conservationists and ornithologists.

been thoroughly debated in Parliament or in public; such debate as there has been has taken place, inadequately, at inquiries into new energy facilities at Windscale, Belvoir and Sizewell, and in the inaccessible pages of academic journals. The best that can be said about the energy debate is that fundamental questions about the overall direction of policy have been raised – lately as much by bodies like the House of Commons Select Committee on Energy as by critics from outside the system. But while decisions are dominated by giant supply interests (public or private), vying to increase their slice of a much diminished cake, it seems unlikely that these questions will receive the attention that they deserve, or that they will be met with satisfactory policy solutions.

FURTHER READING

R. Bending and R. Eden, *UK Energy: Structure, Prospects and Policies*, Cambridge, 1985.

A. Blowers and D. Pepper (eds), *Nuclear Power in Crisis: Politics and Planning for the Nuclear State*, London, 1987.

K. Boyfield, *Put Pits into Profit*, London, 1985.

D.R. Cope and P. James, *Earthing Electricity*, London, 1988.

CPRE (Council for the Protection of Rural England) and FoE, *Electricity for Life*, London, 1988.

A. Glynn, *The Economic Case Against Pit Closures*, Sheffield, 1985.

G. Leach *et al.*, *A Low Energy Strategy for the United Kingdom*, London, 1979.

P.R. Odell, *British Oil Policy: A Radical Alternative*, London, 1980.

8 Land use and landscape

There are two broad types of approach to land use and landscape description. The first is intentionally subjective, relies upon intuition, and deliberately involves the observer in attaching value judgements to the results. In the second approach the observer seeks to be objective and to apply scientific criteria. The intention is to produce results which are based on quantitative information and which minimise personal bias. This chapter deals with the second approach, and attempts to provide an integrated description of the landscapes and land use in Britain today.

A fundamental problem in the interpretation of landscapes is that since they are the product of many interacting factors they vary continuously. Because the elements that make up the landscape combine with each other in complex ways, landscapes appear heterogeneous and many observers respond by simply describing a particular – usually very distinctive – landscape. Moreover, the aesthetic appreciation of scenery, with the complicated psychological aspects of perception, is superimposed upon the direct observations of landscape. Variation within landscape consists of a continuous series of gradients of associated elements – for example an altitudinal gradient with its connected features may dominate at one scale while within a uniform altitudinal zone detailed landscape differences may be generated by soil. The task of landscape description is to divide the continuous variation into classes that are convenient to handle – divisions which are inevitably arbitrary since there is no evidence of discontinuity.

Having determined the classes the next stage should be to attach values to them. Liddle emphasised the importance of this separation, which is often confused in practice, and recognised that three basic processes are involved in landscape assessment: (a) selection of the features to be recorded, (b) recording of the features, and (c) analysis of the records.[1] By examining a series of previous studies he showed the complexity of the operations involved in evaluation. Because landscape description has been used mainly to identify areas which are considered to have a high scenic quality (i.e. aesthetically pleasing land) value judgements are usually involved at an early stage.

An important parallel may be drawn between landscape description and vegetation survey. Vegetation variation is also continuous, with few demonstrable discontinuities, and heterogeneity within species assemblages is a major problem in separating classes (or types). Most observers have therefore relied upon intuition to interpret the complexity observed in the field and have selected visually homogeneous areas for classification purposes. As with landscape analysis, the development of objective methods has proceeded in parallel with the advent of computers, a major objective being to reduce observer bias. However, the botanist is at least able to record species and cover data, whereas landscape features are not easy to quantify, either singly or in combination. Some land uses, such as wheat or barley, are relatively easy to determine but others, such as the different types of moorland vegetation, are more difficult.

The present chapter discusses in further detail the variety of approaches available, both for land use and environment, before describing an integrated system, which is then used to provide a general description of Britain.

METHODS OF CLASSIFICATION

Landscape classification

The appreciation of landscape and the desire to distinguish natural units has gradually developed from Victorian times on a course parallel with increasing leisure and pressure on the countryside.[2] Several government agencies, such as the Ministry of Agriculture and the former Soil Survey of England and Wales, have produced classifications which, although not primarily intended for landscape, contain implicit information about the type of scenery likely to be present.[3] Three methods are frequently noted as being significant in the development of objective approaches to landscape evaluation. Firstly, Linton attempted a direct classification of Scottish 'landform landscapes', based on absolute and relative relief.[4] Gilg then tested the method and concluded that it had potential and was worthy of extension, although the scale of values needed testing outside Scotland.[5] The method was subsequently applied to an appraisal of countryside recreation, and was later incorporated in a slightly modified form in an environmental information system for Scotland.[6] A second method was developed in 1971 in the Midlands by a research group which attempted to process detailed measurements of components of the landscape statistically. This work has, however, been criticised as failing to produce results more efficiently than descriptive methods.[7] The third widely used procedure is based on Tandy's method in which various landscape elements are identified and summarised to produce maps of landscape character.[8]

However, objective methods of landscape evaluation are not widely used and most workers still rely upon subjective evaluation. The majority of users are in development control with local government who are always under pressure and are dealing *ad hoc* with problems. Intuitive methods are convenient and more rapid, hence their widespread use. Strategic studies, however, need a broader base.

General classifications

In Britain, current coverage of environmental attributes is diffuse even though the individual features of the physical environment are reasonably well described. Physiographic information is given in the Ordnance Survey (OS) maps; climatic factors are covered by meteorological maps, geology by maps at a scale of 1:625,000 and soils in England and Wales by maps at a scale of 1:1,000,000.[9] Thus there is a reasonable description, at a low level of resolution, of the basic physical features of the land. However, the pattern of national and regional variation resulting from the association of these features has received little study, although it is this combination which determines the biological and land use characteristics. Selected features of slope, climate and soil have been combined into an assessment of agricultural potential over England and Wales.[10] Such capability classifications are now widely used in many countries but the field of capability interpretation is still developing and the Agricultural Land Classification of England and Wales is being revised.[11]

The Forestry Commission has its own system of assessment of forestry potential, based on climatic and soil characteristics.[12] For agriculture, although maps are not generally available, the environmental criteria associated with yield classes of different crops are reasonably well established and therefore can be derived from maps of basic environmental data. The Commission's agricultural and forestry potential classifications are interpretations of the original physical environment data. Whilst they help in assessing the potential for specific land uses, they are of little help for defining the potential for alternative land uses or different management regimes because they do not retain the independent basic data. However, in contrast to the physical features, data on the biological environment at the regional or national level are very sparse. Ecosystems in particular have been the focus for a great deal of effort, most of it purely descriptive and usually fragmented. No national picture of the biological environment has yet been produced.

An integrated land classification system

In order to improve the existing data base for environmental description at the national level it has proved necessary to adopt a sampling strategy, since cost prohibits complete coverage of any system. Environmental strata are needed, analogous in many respects to the social strata used in opinion polls, so that an adequate framework is available on which to define British landscapes.

The requirement for such a system was recognised in 1975 by the Institute of Terrestrial Ecology and a project was set up to achieve the main objective of a summary of the overall environment of Britain, able to provide a series of environmental strata from which representative samples could be drawn. The techniques had been developed in a survey of the county of Cumbria which in turn developed from an earlier survey of the Lake District.[13]

The project involved the use of computer analyses to define the environment as expressed by a large number of attributes, rather than by a few variables measured in detail. In the original Cumbria survey, the National Grid of 1km squares was used as a sampling basis, in the way quadrats are used to sample vegetation. It was not considered desirable to use spatial units that were initially defined by environmental variables (for example, 'upland' units of landscape), because of the difficulty of maintaining a consistent standard over the whole country and because of the dangers of prejudging the importance of particular environmental variables. The approach is thus very different from the traditional way of defining landscapes. The 1km squares were a convenient arbitrary base for dividing the land surface in order to record the initial data at a scale suitable for subsequent field survey. They also formed a convenient reference system for mapping.

The principle behind the approach is that the major ecological parameters are determined by such factors as altitude which can be recorded from maps and which in turn are correlated with the underlying natural patterns. There are three main phases:

1. Land classification: the recording of environmental data from maps and subsequent analysis to produce a land classification which is used for stratification.

2. Field sampling: the strata (land classes) are used as a basis for field survey of various land use and ecological parameters.

3. Characterisation and prediction: the land classes are then redefined, the distribution of particular features measured, and, finally, land use and ecological features are predicted for areas in which only the land class is known.

A further principle adopted at the outset of the project was that all the data used should be readily available and that, although computers were to be used in analysis of data, all the major stages of the data extraction would need only simple arithmetic. Hence, although mathematical procedures are involved in the data analysis, the application of the results does not require specialist knowledge.

In order to obtain a standard grid of squares, the basic data were extracted from the centre of 15 × 15km intervals, resulting in a total of 1,228 squares over Britain. The main guidelines used in the selection of data features were to include relatively unchanging variables that provided an expression of the underlying environment and not to include many measures of the same factor (e.g. temperature) merely because they were available. The data base for the first phase of the British survey included the following types of information:

1. Climatic: data from climatic maps on a scale of 1:1,000,000 representative of the range of climatic data available.

2. Topographic: data from 1:50,000 OS maps, incorporating features such as altitude and slope.

3. Human artefacts: data from features available from the 1:50,000 OS maps.

4. Geological series: data concerning the presence of the main geological series and surface drift.

The data were originally a mixture of attributes and variables, but the latter were converted into attributes by separation into four equal classes. A total of 282 attributes was eventually used, and 32 land classes were produced by successive division into 2, 4, 8 and 16 divisions, thus representing a progressive dissection of the land surface of Britain.[14] The analysis identifies indicator attributes which may then be used to assign any single kilometre square in Britain to its appropriate land class. Subsequently a further 4,826 squares were assigned to their appropriate land classes, giving in total 6,040 squares in Britain. We can thus obtain rapid estimates of the land class composition of any region in Britain without recourse to further computer analysis. We can also compare the variation in land type between areas, or place the land characteristics of an individual site, for example a potential reservoir or valley, into a regional or national context.

The land classes show well-defined geographical distribution patterns but, as with Cumbria, distinctive unexpected patterns emerge which are interpretable in terms of their combinations of environmental features. The first division is largely associated with altitudinal and climatic features, but in later divisions features such as geology and slope become important in distinguishing land classes.

Descriptions of the land classes

The approach taken was to use the information derived from the classification for the land classes to build up their characteristic features. The complete data are published elsewhere.[15] We have also produced drawings of typical squares to demonstrate the link between landscape land use and land form.[16] Although we have plenty of ground photographs of the sample areas, it has always been difficult to find typical examples containing sufficient features. It was therefore with some trepidation that the Cambridge collection was approached for suitable aerial photographs to illustrate the land classes here. Sixteen core areas were identified where squares of adjacent land classes were likely to be present in sufficient numbers to be recognised in a photograph and reference was made to the collection. It was, however, relatively easy to find oblique aerials at about 1:5,000 that were eminently suitable, as these were appropriate to the scale needed to demonstrate the classes. In fourteen of the sixteen pictures of classes, appropriate areas were found near the core locations identified. These photographs are thus presented in sequence to form the basis of the description of the landscape of Britain. It will be noticed that such an approach plays down the 'unique' areas and relates to 'ordinary' places. The special areas of landscape are well covered in many texts, from National Park Guides to Dudley Stamp's classic work, *The Structure and Scenery of Britain*. However, by definition, such texts cover only exceptional areas and these constitute a small proportion of the whole landscape. The basis of the present methodology is that the whole landscape is treated uniformly, enabling people to fit their own familiar landscapes into context.[17] Because the variation is continuous, no land class is completely homogeneous – rather it contains a range of characteristics within defined bands. The descriptions can be used in a general sense to identify similar landscapes.

The main characteristics of the land classes in Britain are shown in the series of photographs that follows and the descriptions are grouped in pairs. Each of the descriptions is in the same format: a brief statement of salient features is followed by an account of land form, and progresses by way of topography and landscape, to soils, land use and natural vegetation. These descriptions are based on the sample described above but currently the whole of Britain is being classified. Detailed distributions and minor modifications will thus be available in due course.

APPLICATIONS

The main advantage of the system of land classification upon which this description of Britain is based is that it is able to treat the land as an integrated system within which the various elements can be compared. The other systems referred to in the introduction are all developed for specific purposes, for example agricultural production. This final section therefore presents several examples of projects which demonstrate the application of the methodology to integrated studies.

Since this chapter was first drafted a series of wider applications has been carried out. However, those described below are useful to demonstrate the technique. The first example was developed during a study known as the Land Availability Study (LAS) commissioned by the Department of Energy. The LAS was designed to estimate the amount of land which might become available for growing trees for energy under various economic assumptions. The resultant Land Availability Model (LAM) provided a means for predicting change in land use at the site-specific level and shows how integrated studies may be achieved.[18]

The core of the LAM lay in the land classification system described above. Potential uses for land in the sample squares can be postulated and the resultant figures for Britain estimated. This mechanism forms the basis of the LAM, which has an advantage over other models in that it provides information on areas of new land use, as well as on the types of land use which would be lost. An assessment of the environmental impact of various practices can therefore be made, showing the way the system can be used to assess interactions. The criterion for predicting change in land use in the LAS was financial performance, calculated on net present value (NPV) of land use over sixty years, at a given discount rate. Economic data for 1977 were used.

The LAM enabled the comparison of the relative financial performance of different, actual and postulated, land uses for individual areas of land in the sample squares and provided estimates for Britain of the areas of new forestry for energy, the potential production and the likely species, thus indicating possible future developments. In addition, the area and nature of displaced land use categories were given, showing that the implications of potential change can be predicted in terms of their associated habitats.

To obtain an estimate of financial performance, a series of some 140 sets of economic values was produced in order to include a range of management systems and yields for forestry. The management systems were: conventional forestry for timber; forestry for energy and timber; and single stem and coppice energy plantations. Sets of NPVs were calculated for each model for a range of discount rates, timber values and assumed values of wood for fuel. Current British agricultural practice was described by, and classified into, some forty production systems. Investigations showed that it was not possible to allocate fixed costs to particular enterprises so, for purposes of comparison, the NPV of the enterprise gross margin over sixty years was taken. The constraints arising from national planning controls relating, for example, to nature reserves, public pressures, and legal impediments to change in land use, were included. The significance of such constraints was noted on each of the sample squares. Their probability of restricting forestry was used in the model, but could be applied equally to changes in agricultural practice.

The LAM enabled examination of the effects of a wide range of economic assumptions on the areas of land predicted as being available for forestry. One example assumed the need for wood for the production of pipeline gas, increasing energy prices, constant timber and agricultural costs and revenues, and a 5 per cent discount rate. Here some 4.6M ha of land were predicted to change to forestry (all with an energy component) with a potential annual production of some 38M dry tonnes of wood for energy and 28M m^3 of timber. However, when constraints were taken into account, some 1.8M ha of land were estimated to change to forestry, producing some 16M dry tonnes/yr of wood for energy and 11M m^3/yr of timber.[19]

The incorporation of nature conservation constraints in the LAM demonstrates the way in which conflicts between economic development and wildlife can be assessed. The system of comparing potential and actual uses for units of land can be readily modified to examine the likely future of wildlife habitats such as small woodlands or herb-rich grasslands. Similarly, the effects of possible changes in the countryside, such as canalisation of rivers, can be examined at a strategic level. The economic implications of restricting agricultural improvements can also be assessed in comparative terms. Further, the examination of the loss of agricultural land to wood energy plantations demonstrates the way in which impacts can be predicted. Such an approach has been used by the Highland Regional Council to examine conflicts between planning constraints, nature conservation, agricultural improvement, and red deer production.[20] As the demand for land changes, so such studies are progressively required.

Initially, each model reproduced the current pattern of land use on the farm, establishing the status quo. Next, the forestry (in terms of area by management system, species and yield) predicted by the LAS was inserted into the farm land use pattern, replacing any existing land use which proved less financially viable. The model was then run to examine the effects of such a change in land use on the on-farm management factors. In broad terms, the exercise confirmed that the area of forestry suggested by the LAS could be accommodated on farms, providing that certain 'key' resources such as capital were available. The size of the resource required varies with farm type. If larger areas of forestry than those predicted by the LAS are allowed, then economies of scale start to take effect. In some circumstances, farm income may be increased by increasing the area under forestry. As with the previous model concerning forestry objectives, a comparable approach could be used for conservation, with wildlife habitats being designated as 'key' resources.

The second example developed at the Institute of Terrestrial Ecology demonstrates the wide application of the Merlewood land classification system, using land classes to examine the impact of changing patterns of agriculture and forestry on rural land use.[21] The implications of several different scenarios were applied directly to categories of land use to assess the general changes that might result. Possible scenarios considered were:

1. Dairying becomes less profitable.
   Grasslands within enclosed land are likely to be mainly affected, as opposed to the hill land in scenarios (3) and (4).

2. Cereals become less profitable.
   These changes are considered to apply mainly to the area of tillage.

3. Hill subsidies are reduced.
   This reduction will lead to a decline in agricultural use of the uplands and would involve subtle ecological effects.

4. Hill and upland sheep become more profitable.

This scenario is the reverse of (3) and would result in a general increase in the use of upland vegetation, with associated effects on its ecology.

5. Confidence in, and support for, forestry increase.
   Changes within grassland categories were not considered likely, in comparison with the other scenarios.

The majority of transfers between the land use categories for the different scenarios were in the order of 5–10 per cent. The loss to urban development, although significant in the long term, is below the level of change that can be considered by this model. The results showed some categories changing considerably because some trends were toward a particular land use, for example coniferous woodland. In other examples, feedback mechanisms are in operation, buffering change in land use as described by Best.[22] For example, in scenario (1), the loss of high-yielding pastures to barley is compensated by the change of poorer pastures into good quality sward. It is only when a major land use category accumulates inputs from other uses that major changes occur. The complexity of feedback and the preliminary nature of the study indicate that more developed research is required to obtain harder information on transfers between land uses. However, it is necessary to consider whether comparable patterns are followed in habitat change, as some evidence suggests that change can take place very quickly. Even so, the transfers in the main crops are equivalent to the changes in recent years indicated by the Ministry of Agriculture, Fisheries and Food (MAFF),[23] and the change in forest area is comparable with that which has taken place over the last ten years, suggesting that the projected changes are reasonable.

Most of the land uses incorporated in the above example were selected to identify implications for associated semi-natural habitats. For example, the changes in scenario (3) directly affect the composition of grassland of great ecological interest. A comparable study could equally be developed by identifying specific habitats such as hedgerows and streams, and examining the implications of changes in rural land use for their occurrence, to assess potential national changes and critical influences on conservation. The effects of stopping straw burning could, for instance, be compared with those of peatland drainage. As the proportions of the different categories of land use vary through the land classes, the system can also be used to examine regional differences. Thus, in scenario (3), the effect on scrub is felt mainly in the downs of south-east England and in the uplands of Wales, Cumbria and Scotland.

The third case study is provided by a recent project on the countryside implications for England and Wales of possible changes in the Common Agricultural Policy. The aim of the project was, first, to identify the policy options available to EC policy-makers; second, to estimate the agricultural production consequences of these options; and, third, to examine the likely impact on the rural environment.

Previously, such studies have been based largely on intuitive judgement, but the availability of the land classification system and an econometric model at Newcastle University enabled the research team to produce a composite model which has been termed the Reading model. The model makes it possible to link changes in commodity prices, determined through the scenarios, with actual changes on the ground. For example, the imposition of a quota on wheat could be linked to the areas in the country most affected. Such changes have important implications for the landscape as changing farming practices alter the appearance of the countryside through the loss of features such as hedgerows and trees.

The first part of the project involved the use of an econometric model to examine the interrelationships between the prices of agricultural products in the EC. The second part used the land classification system described in the present chapter to gauge the agricultural impact on chosen areas of Britain. The two models were linked by a computer programme which enabled price changes to be tied to the areas of land most likely to be affected. These changes were then linked to possible associated changes in landscape character. For example, if quotas were introduced for wheat, the area of land used for wheat would decline and be replaced, probably with grass, with some benefit to landscape. However, this decline would not occur, as might be expected, in the central East Anglian production area because here the land is optimal for wheat production. Rather, changes may take place in regions such as the West Country, where cereal enterprises are marginal in economic terms. Even so, effects on the landscape would be significant in this case, largely because the agricultural intensification associated with arable expansion is not likely to take place. The retention of traditional systems of grazing by beef cattle and lowland sheep leads to maintenance of traditional landscape patterns.

Strategic studies such as these are today essential for planners, who need to be aware of the modifications which can take place in the landscape. Although landforms remain unaltered, the landscape itself is changing continuously and must be seen as a dynamic system. Changes are now very rapid indeed and if the planning process does not take account of potential scenarios, the policies cannot be adjusted accordingly. The primary objective of the methodology described in the present chapter is to aid the interpretation of such variability through an understanding of the interaction between the various elements. Whilst it is a truism to say that the various elements depend on each other, few other approaches have recognised that dependence and used it in order to provide an overall assessment of the landscape.

The photographs that follow have been chosen to be representative of the Land Classes, though not all features will be present in all cases.

118 Land Class 1: undulating country with varied agriculture, mainly grassland; and Land Class 2: open, gentle slopes, lowland with varied agriculture.

These Land Classes represent many of the truly lowland landscapes of southern England and contain a considerable range of local variation, for example in architectural styles of the buildings and in local land use reflecting local soil types. They do, however, have a consistent land form, involving a range of variations within the broad definition of alluvial plains – as seen on the left of the photograph. Lowland landscapes tend to be more difficult to define than their more varied upland counterparts and, moreover, need more detailed recording of specific features, e.g. field size, to enable finer divisions to be made. The two Classes tend to be more intermixed than many of the upland Classes, depending very much on small-scale patterns in the topography.

The land forms of Land Class 1 are mainly level alluvial plains or slightly rising land along the margins of major river valleys such as the Severn and Avon. The land is usually flat or at most gently rolling, and, by lowlands standards, the landscapes are varied. Hedges are prominent and there are trees and scattered farm buildings. Small hamlets and villages are frequent, with larger towns also present. The land use is dominated by grassland, both short-term and permanent, but with significant areas of cereals and horticulture. Native vegetation is mainly restricted to rough grassland on slopes difficult to cultivate, though lowland heaths are also important features on appropriate geological strata.

In contrast, Land Class 2, as on the right of the photograph, is characteristic of the downland summits and scarps, or low ridges of the Weald, for example, with generally sweeping curves and smooth slopes at low altitudes. The landscape is mainly open and wooded downland with few hedges and trees, although locally there are extensive areas of woodland on poorer soils. Grassland, particularly temporary, is important but cereals and other crops are also significant. Scattered farmhouses form the most typical buildings but, in terms of area, towns are important and overall there is a high proportion of built-up land. Native vegetation is mainly confined to rough grassland typically on unimproved chalk slopes; but bracken or some lowland heaths are also present.

119 Land Class 3 : flat arable land, mainly cereals with little native vegetation ; and Land Class 4 : flat intensive agriculture otherwise mainly built-up. Land Classes 3 and 4 dominate East Anglia but show rather different patterns of detailed distribution – the former dominates the central area of the region and margins of the Weald, whereas the latter occurs towards the margins of both areas.

Land Class 4 occupies the plains of East Anglia, or very small river valleys and the broad areas between them. The land is level, or nearly so, and almost all is at low altitude. The landscape consists of prairie-type lowlands where highly intensive agriculture has removed many of the hedges and trees. The land use is dominated by cereals, although other crops are also important, and short-term grassland, of species such as rye grass, still remains in significant local patches. In Land Class 4 urban development is usually in the form of small hamlets or villages, rather than major conurbations. The foreground is highly built-up Land Class 4, with trees around the fringe of the town.

By contrast the background, with Land Class 3 predominant, has a patchwork of arable land uses with relatively little urban development. Native vegetation is virtually absent but, where present, is usually in small patches of residual old pastures or meadows. Land Class 4 is typically fenland or flood plains with intricate drainage patterns and is still liable to flooding. The landscape consists of intensively farmed lowlands, often with ditches, and with few hedges or trees, except around built-up areas. Cereals, crops and horticulture share the agricultural land but urban areas, particularly London, dominate much of the land and are the principal land cover. Some good short-term grassland still remains however, but there is very little woodland. Native vegetation is virtually absent, but, where present, consists mainly of old pastures.

120 Land Class 5: lowland somewhat enclosed land, varied agriculture and vegetation; and Land Class 6: gently rolling enclosed country, mainly fertile pastures.

Land Class 5 covers a wide area, with a dispersed distribution in southern Britain. It has a rather varied land form, from gentle scarplands to downland and valley floors, but with steeper slopes than the previous classes. By contrast, Land Class 6 is more restricted: it is particularly typical of the tablelands of the south west, with plateaux dissected by many small rivers. These features are exemplified in the photograph, with Land Class 5 being the open rolling country of the foreground and Land Class 6 occupying the more complex land of the background. Land Class 5 is more closely related to Land Class 2; with many lowland features such as hedgerow trees and small woodlands. The landscape of Land Class 5 is mixed farmland with a variety of different crops, although grassland predominates. Overall, there is much urban development ranging from vernacular farm buildings to hamlets, small villages and towns. The native vegetation is limited in extent but, where present, it is varied, including patches of marshland and areas of bracken and rough grassland, often in places that agricultural improvement has not reached.

Land Class 6 represents the familiar Devon landscapes, also typical of south Wales, with complex topography and many broad, even slopes and the majority of land at medium and low altitudes. The landscape is intricate, with small fields often enclosed by hedges and banks, and many trees, both along hedge lines and in small copses. The land use is dominated by grassland, mainly permanent than temporary, as in previous classes. However, barley is also an important crop for animal feedstuff. Native vegetation is mainly associated with nutrient-rich soil, present typically in poorly drained field corners, though around the margins of the hills some small areas of grassland with more upland species may be found.

121 Land Class 7: coastal, with variable morphology and vegetation; and Land Class 8: coastal often estuarine, mainly pasture, often built-up.

The Classes follow the coastline and serve to separate coastal squares from the inland squares. However, they inevitably cover a wide range of variation, because the classification in use here was not set up specifically to study the coastline. The general distinction between the two Classes is clear though, with the foreground typical of the generally estuarine Land Class 8, while the background shows the more variable Land Class 7, with coastal cliffs often cut back into tablelands. In Land Class 7 the land use is variable and depends upon exposure and soil type, with brown earths predominating. The main use is pasture land with some arable, typically barley. The native vegetation is quite extensive on slopes that are too steep to cultivate. There is a range of types, from coastal heath to bracken, and even moorland on exposed areas. Land Class 8 is more typical of the marine alluvial plain bordering estuaries and is mainly flat, although some steeper areas may also be present. The landscape is typically flat and windswept, backed by good farmland, but is often dominated by urban development. The land use is mainly permanent grass, but usually there is also arable land around the edges of extensive mudflats. Seaside towns are a frequent feature and dominate many areas. Native vegetation is often limited to saltmarsh, but small neglected areas may also have marshland with rushes or bracken.

122 Land Class 9: fairly flat, open intensive agriculture often built-up; and Land Class 10: flat plain with intensive farming, often arable grass mixtures.

These Classes belong to the midland and north-east plains and are closely interrelated, in the same way as are Classes 1 and 2 and for similar reasons. In both cases the landscapes have few strongly distinguishing features. Land Class 9 is in the foreground occupying more open lowland country, in contrast to Land Class 10 in the background with more trees, hedgerows and woodland. Land Class 9 occupies mainly the valley floors and flood plains of the large rivers of the Midlands and is rolling or almost flat country at medium to low altitudes. The country is usually open, with intensive agriculture often leading to declining hedgerows and loss of trees. Short-term grassland dominates the land use but there is also a high proportion of barley. Urban areas are also frequent, particularly the margins of towns and large villages. Native vegetation is very limited with most of the land being farmed or built-up. Land Class 10 is usually less flat, with long gentle slopes present. The landscape, although intensively managed, typically has many hedgerows, trees and small woods. The land use is dominated by a mixture of permanent and temporary grassland and a similar proportion of barley and wheat. As with Land Class 9, native vegetation is very restricted.

123 Land Class 11: rich alluvial plains, mainly open with arable or pasture; and Land Class 12: very fertile coastal plains with productive crops.

Land Class 11 has a restricted distribution inland and south-east from the coast, whereas Land Class 12 is tightly clustered to the east and is more typical of fenland. Land Class 11 includes alluvial plains or low, broad ridges drained by small streams, with very gradual slopes or flat land. The landscapes are very open with large fields and relatively few hedgerows. Arable land predominates, particularly wheat and barley, but there is also an appreciable amount of temporary grassland. Urban development is also an important land use, with small towns and villages being particularly common. Native vegetation is very restricted, and occurs mainly in old water meadows or field corners beside the large rivers or dykes. Land Class 12 consists mainly of the fens or adjoining flood plain, with flat or virtually flat land almost at sea level. The landscapes are prairies with derelict hedges, few trees, and boundaries that are typically formed by drains. Arable land is the dominant land use, mainly wheat and barley, but also other crops and some horticulture. As with Land Class 11, there is considerable pressure for development, leading to extensive allocation of land for urban use. Native vegetation is very restricted because of the high value of the land and there are few neglected areas, in field corners or elsewhere. Both Classes involve highly intensive agriculture, with associated eutrophication of rivers, and high nitrogen levels in particular.

124 Land Class 13: somewhat variable land forms, mainly flat, heterogeneous land use; and Land Class 14: level coastal plains with arable, otherwise often urbanised.

These Land Classes contrast strongly, with Land Class 13 inland (towards the back of the print) and predominantly agricultural, and Land Class 14 on the coastal plain in the foreground, with urban development being the principal land use. Land Class 13 is rather heterogeneous in terms of land form, varying from low ridges on the margins of the coastal plains to scarps and river valleys further inland. The slopes are generally smooth but can occasionally be quite steep. The landscapes present are rather varied, typically with hedged small fields and trees, although in the north of England these tend to be absent. The variability is reflected in the land use which, although dominated by leys and permanent grass, also features barley and wheat, and even some areas of open range grassland in northern areas. The areas of native vegetation also vary, with rushes and bracken throughout, and moorland and peatland in northern localities as well. Land Class 14 occupies mainly marine or alluvial flood plains bordering estuaries, predominantly in northern England but extending into southern Scotland. The land is flat and tends to be quite open. Fences or neglected hedges typify landscapes which are much affected by adjacent urban development. The land use is mainly arable, but temporary grassland is also important. A major feature is the extensive urban development, dominated by major coastal conurbations such as Blackpool. There is very little native vegetation. This Land Class, although an important one in the British context, and one clearly separated from its southern counterparts, is also one which is often overlooked.

125 Land Class 15: valley bottoms with mixed agriculture, predominantly pastoral; and Land Class 16: undulating lowlands, variable agriculture and native vegetation.

Land Class 16 is in the foreground with more open, rolling countryside than the typical narrow valley sides and scarps of Land Class 15, so typical of the northern Welsh valleys. The land forms of Land Class 15 include elements of low hills, scarps and valley bottoms as well as valley floors. The topography is thus complex, with shallow slopes (although they can occasionally be steep) and flat land, usually at low, but extending to medium, altitudes. The landscapes are intricate, with many natural features reflecting the complexity described above – small woods, rough slopes and groups of trees are typical. The land use is mainly a mixture of permanent and temporary grass but there are considerable areas of barley and some land with other crops. Villages and quite extensive woodland also contribute significantly to the overall land use pattern. Native vegetation is restricted, with bracken the most widespread type on steep banks, but otherwise only small marshes and rushes in poorly drained corners. Land Class 16 is more typical of flood plains or valley floors in low hills – gently undulating, often incorporating level areas. The landscapes are variable lowland and are well tended, with many hedges and trees and with larger fields than Land Class 15. The land use is more varied because of its more northerly distribution, although permanent and temporary grass still predominate, with an admixture of arable land, particularly cereals. Otherwise there is a range of uses from rough pasture to some areas of open grassland. Urban development is limited mainly to small hamlets and villages, in addition to the usual scattered farmhouses. Native vegetation is only found in small areas, but it occurs quite frequently for a lowland environment. It varies from bracken to rushes and also includes some small patches of peatland and moorland.

126 Land Class 17: rounded intermediate slopes, mainly improvable permanent pasture; and Land Class 18: rounded hills with some steep slopes, varied moorlands.

The photograph shows clearly the difference between the two Classes. Land Class 17 is mainly the enclosed land in the valley margins, with some open land above, and Land Class 18 comprises the higher mountain slopes with more rugged terrain. Land Class 18 has a wider geographical range including some areas in the west of Scotland. Land Class 17 is generally formed of plateaux or tablelands with scarps, often dissected by small rivers. There are some gentle slopes, but mainly quite steep hillsides at medium altitudes. The landscape is a mixture of enclosed and open land, usually with walls and open hillsides, and so forms classic marginal upland.

The land use is mainly permanent pasture, although there has been much improvement. There are, however, still some areas of woodland, both broadleaved and conifer. There is little built-up land. The extent of native vegetation often depends upon the ease of tractor access, and consists mainly of grassland types, with some bracken and rushes. Land Class 18 is more rugged, with glaciated river valleys and steep scarps, backing on to tablelands in the south or distinct mountains in the north. The landscapes are mainly open mountain hillsides, but include some lower land bounded by the mountain wall, which has probably been enclosed only since the nineteenth century. The land use is predominantly rough grazing with open range sheep country and a small amount of improved grassland, at the lower levels. The native vegetation is mainly moorland and mountain grassland with some peatland and limited areas or bracken and rushes.

127 Land Class 19: smooth hills, mainly heather moors, often afforested; and Land Class 20: mid-valley slopes, wide range of vegetation types.

The boundary between these types is shown clearly here, though the valley bottom to the left belongs to a different Class (probably Class 9). Land Class 20 occupies the steep scarp, including the boundary between the enclosed land and moorland, in contrast to Land Class 19, which typically comprises the plateau slopes. The latter therefore occupies the broad ridges or flat-topped rounded summits with small rivers. The topography is mainly moderately steep slopes, but there are also some rather steep hillsides at the higher altitudinal limits of the class. The landscape is predominantly moorland, in which ling heather plays a dominant role, with a lesser proportion of rough and mountain grassland and some permanent grass at lower levels. There are often grouse present but the land is also highly suitable for afforestation which has been occurring in some areas. Woodland, invariably conifer, thus also forms an important land use cover. Native vegetation is widespread, except where improvement has taken place or where conifers have been planted. In general, ling heather is the most dominant species, but many other acid grassland species are also important. Land Class 20 occupies upland river valleys or mid-valley hillsides on the margins of upland areas. The topography is often complex, and includes steep hillsides and more moderate gradients at medium to high altitudes. The landscape is a mixture of upland and marginal lowland, with fences and walls, a few scattered trees, and buildings dotted throughout. The main land use is permanent pasture, although there is also much improved grassland at the lower levels. Otherwise, it is largely open range native vegetation, mainly poor acidic grassland and moorland, with some peatland at higher elevations on more poorly drained sites.

128 Land Class 21: upper valleys and rocky outcrops and bogs; and Land Class 22: margins of the high mountains, generally moorlands, often afforested.

The general distinction between these classes is visible here, with Land Class 21 occupying the valley sides and Class 22 the open country behind. Land Class 21 generally occupies peneplain surfaces, with complex drainage patterns, or broad ridges with indistinct summits. The topography consists of mainly steep hillsides, with some minor, moderate slopes towards the valley bottoms. Typically, the landscape is bleak moorland with few relieving features. Although walls and fences may enclose land at lower levels, these are often in bad condition. The land use is open range grazing with sheep or deer and extensive coniferous forest. Native vegetation covers most of the land surface, with ling heather moorland as the main type, although peatland with deergrass is also widespread in more northern or poorly drained situations. Land Class 22 covers a wider geographical range and conditions, and occupies the dip slopes of plateaux or broad glacial valleys leading to rounded mountain summits. The topography consists of variable gradients from steep to moderate, although the latter predominates. The landscapes are typical of the high moors of the Pennines and the margins of the high hills in northern Scotland, being of mainly open moors with occasional large blocks of woodland. The land use is mainly rough grazing and open range sheep and deer are widely present. At lower elevations there is some enclosed land with improved pasture and even some fodder crops in sheltered situations. This Land Class has a high potential for afforestation, and much planting is currently taking place. However, native vegetation still occupies the majority of the land – mainly moorland and peatland, with ling heather being by far the most important species.

129 Land Class 23: high mountain summits with well-drained moorlands, typically in the east; and Land Class 24: upper steep mountain slopes of the west, usually bog covered. Land Class 23 occupies in general the summits of the hills, whereas Class 24 comprises the slopes; and there is a tendency for the former to be in the east and the latter in the west.

These Classes form the main mountain summits in Britain. Because of their characteristic forms they are easily recognised as different from other mountain groups. Therefore Land Class 23 includes the Cairngorms and the Mamores whereas Class 24 is more typical of mountains such as The Five Sisters of Kintail and An Teallach in Wester Ross. For the present purpose, however, they have sufficient in common to be treated together. Land Class 23 is formed of ridges, scarps and corries leading to mountain summits or glaciated valleys. The topography consists mainly of very steep hillsides at high altitudes, with gentler slopes much less widespread. This Class includes many of the well-known mountainous landscapes of Britain, with wide vistas and spectacular views. The land use is open range grazing, usually for deer, but there are also sheep in the south. The native vegetation is mainly of moorland types (with ling heather predominant), but also some mountain grassland and peatland types.

In the west of the country Land Class 24 is more typical of valley sides rather than summits. It often reaches from the valley floor to the upper slopes. The topography therefore consists of precipitous and extremely steep slopes, with many cliffs, rocky outcrops and classic mountain features. The landscapes are often spectacular, with rugged mountain slopes, fast-flowing streams and waterfalls. As with Land Class 23 there is open range grazing on this poor land, in this case probably of deer but with some sheep on the lower slopes. The native vegetation also is mainly of peatland, because of the high rainfall in the west, with purple moorgrass and deergrass predominating. Ling heather is also important and a variety of other species, for example bracken, may be found at lower elevations on better-drained scarps.

130 Land Class 25: lowlands of northern England and Scotland with variable land use but mainly arable; and Land Class 26: fertile lowlands of northern England and Scotland with intensive agriculture.

The overall land forms are similar here, though the large fields of the left foreground are more typical of Land Class 25, whereas Class 26 occupies the more variable land behind. Land Class 25 occupies alluvial flood plains, in areas such as Fife and inland of Aberdeen, and valleys of largely glacial origin, with virtually flat topography at low altitudes. The landscape is usually of intensively farmed lowlands in Scotland and northern England, which differ from their southern counterparts in the small extent of hedges and the predominance of wire fences or walls. The land often has a windswept appearance and there are scattered farmhouses. The land use is dominated by barley, but there is also a high proportion of temporary grassland, though little permanent grass. Native vegetation is thus very restricted, consisting only of a few damp field corners and meadows. Land Class 26 is a little more variable, occupying valley floors and coastal plains of glacial origin, such as the Black Isle and the central Scottish valley, but sometimes with outcrops of other material. In general the land is fairly flat. The landscape is of a rather mixed lowland character, rather open in comparison with the southern lowland landscapes, and with walls and fences instead of hedges. Many areas are affected by urban development. The land use is dominated by grassland – particularly short-term leys – but there is also an appreciable acreage of barley. Urban areas cover a significant proportion, with small towns being important as well as scattered villages. Native vegetation is limited in extent but variable in its composition, and is due to outcrops of underlying acidic rocks emerging through the mantle of glacial till. It thus varies from marshland through to bracken to purple moor grass and deer grass.

131 Land Class 27: fertile lowland margins with mixed agriculture; and Land Class 28: varied lowland margins with heterogeneous land use.

These Classes occur inland, in major river valleys, or in lowlands on the margins of upland areas, such as the Lomond hills or the Trossachs. Land Class 27 is typical of the foreground of the photograph, whereas Class 28, on the more variable margin of the hills behind, is influenced by the harsher conditions. The land forms differ but are mainly valley floors, with low ridges between them. The topography varies, from mixtures of slopes to gentle gradients, mainly at medium or low altitudes. The landscape is generally of well-tended, fenced lowlands with vernacular farm buildings and a few scattered trees – of a more obvious northern appearance than the softer southern Land Classes. The land use is dominated by grassland, particularly permanent grass, but the land is sufficiently good to carry a significant area of barley. Small woodlands are also scattered through the landscape, mainly of conifers but with some hardwoods. Native vegetation is restricted in extent and varies very much according to the conditions under which it occurs, whether marshland or more upland moorland types in exposed situations where poor acid rocks are emergent. Land Class 28 is very heterogeneous because the land forms in the southern part of its distribution give rise to quite different conditions from in the north in Sutherland. In the south, as shown here, the level alluvial plains give rise to fertile conditions in which the land is relatively flat. Hence the lands are well tended, with enclosed fields, dominated by intensively managed short-term grassland but with some permanent grass and even some barley in sheltered situations. There is relatively little native vegetation, with that present being typical of northern agricultural areas, e.g. marshland and rough grassland. By contrast, the land with similar topography in the north has very different vegetation because of the exposure and very poor underlying rock. Here are found windswept moorlands and peatlands with low grazing pressure from open range sheep and deer. Although currently there is little woodland, the low cost of the land is leading to its acquisition by the big forestry companies and the beginning of large new conifer plantations.

132 Land Class 29: sheltered coasts with varied land use, often crofting; and Land Class 30: exposed coasts dominated by bogs.

The land around the highly indented sea lochs belongs to Land Class 29 and the exposed outer coasts are included in Class 30. The west coast of Scotland is separated into these two Classes, with the inner land along the loch sides belonging to the former. The very different exposure experienced by the two Classes is reflected in the abundance of trees. Land Class 29 is composed of classic indented coastlines, with wave-cut platforms and superb sheltered beaches. The topography is very complex, usually with easy slopes but with some steeper mountain sides, sometimes reaching medium altitudes. The landscape is very variable and is generally regarded as among the most scenically attractive in Britain, with many contrasting elements present, e.g. cliffs, coves and small beaches. The land use is mainly open grazing, with the crofting communities having small areas of tillage and hay around the steddings. There is some built-up land, with small towns and villages, at the heads of the lochs. Although there is a relatively small area of woodland, it is an important feature in the landscape, as are the areas of open water. The native vegetation is extremely variable and, although mainly consisting of moorland and peatland types, there are many other elements present, such as marshland and some sand dunes. Land Class 30, by contrast, is very exposed and often consists of bare, eroded, ancient peneplains with meandering streams and low hills. The topography is variable, from almost flat to complex, usually at low altitudes, though sometimes extending a little higher. The landscapes are dominated by the proximity of the sea, with open moorland and many rock outcrops and lochs. The native vegetation is mainly peatland, with some moorland and mountain grass, but virtually no cultivated land. As such, it is a declining landscape, with abundant evidence of old crofts. The land is now mainly used for open sheep grazing. The exposure means that there is little woodland, with small potential for expansion.

133 Land Class 31: cold, exposed coasts with variable land and crofting; and Land Class 32: windswept low hills, covered with bogs.

The distinction is clear here between the coastal margins, with scattered settlements and agricultural land, as opposed to the largely bog covered heathland, so typical of Shetland. Land Class 31 thus consists of indented drowned coastlines, with small coastal plains backed by low hills. The topography consists of gentle slopes with some steep cliffs where marine erosion has cut back into the land surface. The landscapes are typically extremely windswept, exposed coasts with strong contrasts present between them and the small, intensively tended fields of the crofts. Overall the land is mainly open range rough grazing, but there is some good land and pasture around sheltered bays and on better soils. Settlements are usually found in these more favoured areas, although overall both classes have the lowest density of urban development in the lowlands of Britain. The native vegetation is widespread away from the agricultural areas. Moorland predominates but, depending upon the drainage and soil conditions, there is some mountain grassland and peatland.

Land Class 32 occupies the peneplain surfaces of the north of Scotland and the Shetlands and has variable topography, at a small scale, with generally smooth slopes at low to medium altitudes. The landscape is characterised by some of the bleakest moorlands in Britain, with wide open spaces, scattered small lochs and eroding peat hags. The land use is mainly very low level open range grazing, principally of sheep, but there are some areas of rough pasture on the steeper slopes. The whole area is almost completely covered by native vegetation, mainly of peatland with much purple moorgrass, cotton grass, deergrass and ling heather. Locally, by streams, somewhat richer marshland has developed.

FURTHER READING

R.G.H. Bunce and D.C. Howard, *Species Dispersal in Agricultural Habitats*, London, 1990.

R.G.H. Bunce and R.S. Smith, *An Ecological Survey of Cumbria*, Kendal, 1978.

P. Greig-Smith, *Quantitative Plant Ecology*, London, 1984.

R.D. Roberts and T.M. Roberts, *Planning and Ecology*, London, 1984.

D.G. Robinson, I.C. Laurie, J.F. Wager and A.C. Traill, *Landscape Evaluation: The Landscape Evaluation Research Project 1970–1975*, Manchester, 1976.

M. Shoard, *The Theft of the Countryside*, Hounslow, 1980.

Notes

Introduction

1 Lord Esher, 'Contemporary planning', in J. K. S. St Joseph (ed.), *The Uses of Air Photography*, London, 1977.

2 J. B. Harley, 'Maps, knowledge and power', in D. Cosgrove and S. Daniels (eds), *The Iconography of Landscape*, Cambridge, 1988.

3 E. Moessard, 'Telephotography from aeroplanes and its application in national defence', *British Journal of Photography*, 59 (21 June 1912), 478–9.

4 D. R. Stoddart, 'The foundations of geography at Cambridge', *Geographical Journal*, 141 (1975), 216–39. Reprinted in *On Geography and its History*, London, 1986, and B. Hudson, 'The New Geography and the new imperialism, 1870–1918, *Antipode*, 9, 2 (1977), 12–19.

5 W. G. V. Balchin, 'United Kingdom geographers in the Second World War', *Geographical Journal*, 153 (1987), 159–80.

6 N. Chomsky, *At War with Asia*, London, 1969.

7 B. Wisner, 'Geography: war or peace studies?', *Antipode*, 18, 2 (1986), 212–17.

8 M. Foucault, *Power/Knowledge: Selected Interviews and other Writings 1972–1977*, C. Gordon (ed.), Brighton, 1980.

9 K. Hudson, *Industrial History from the Air*, Cambridge, 1984.

10 C. Lewis, *Sagittarius Rising*, London, 1936.

11 D. Charlwood, *No Moon Tonight*, Sydney, 1956, and London, 1984.

12 Chomsky, *At War with Asia*.

13 Harley, 'Maps, knowledge and power'.

14 T. Williamson and L. Bellamy, *Property and Landscape: A Social History of Land Ownership and the English Countryside*, London, 1987.

15 R. Banham, *Megastructure: Urban Futures of the Recent Past*, London, 1976.

16 H.R.H. Prince of Wales, *A Vision of Britain: A Personal View of Architecture*, London, 1989.

17 J. Thomson, *Illustrations of China and its People*, 4 vols, London, 1873.

18 I. Jeffrey, *Photography: A Concise History*, London, 1981.

19 J. Tagg, *The Burden of Representation. Essays on Photographies and Histories*, Basingstoke, 1988.

20 M. Foucault, *Discipline and Punish. Birth of the Prison*, trans. A. Shedidan, Harmondsworth, 1977.

21 Tagg, *The Burden of Representation*.

22 B. Smith, *European Vision and the South Pacific*, 2nd edn, New Haven, London, 1985.

23 Lady E. Eastlake, 'Photography', *Quarterly Review* (London) 101, (April 1857), 442–68. Reprinted in B. Newhall (ed.), *Photography: Essays and Images*, London,

24 A. Scharf, *Art and Photography*, Harmondsworth, 1968.

Chapter 1

1 A. MacEwen and M. MacEwen, *National Parks: Conservation or Cosmetics?*, London, 1982; A. MacEwen and M. MacEwen, *Greenprints for the Countryside?*, London, 1987.

2 W. M. Adams, *Nature's Place: Conservation Sites and Countryside Change*, London, 1986.

3 D. F. Ball, J. Dale, J. Sheail and O. W. Heal, *Vegetation Change in Upland Landscapes*, Cambridge, 1982.

4 Countryside Commission, *A Better Future for the Uplands*, Cheltenham, 1984.

5 D. A. Ratcliffe, *Highland Flora*, Inverness, 1977.

6 J. Raven and M. Walters, *Mountain Flowers*, London, 1956.

7 A. S. Watt and E. W. Jones, 'The ecology of the Cairngorms, I: the environment and the altitudinal zonation of vegetation', *Journal of Ecology*, 36 (1948), 283–304.

8 O. W. Heal and D. F. Perkins (eds), *Production Ecology of British Moors and Montane Grasslands*, New York, 1978.

9 R. A. H. Smith and G. I. Forrest, 'Field estimates of primary production', in O. W. Heal and D. F. Perkins (eds), *Production Ecology of British Moors and Montane Grasslands*, New York, 1978.

10 C. F. Summers, 'Production in montane dwarf shrub communities', in Heal and Perkins (eds), *Production Ecology of British Moors and Montane Grasslands.*

11 H. Godwin, *The History of the British Flora: A Factual Basis for Phytogeography*, Cambridge, 1975; H. Godwin, *The Archive of the Peat Bogs*, Cambridge, 1975.

12 M. J. C. Walter and J. J. Lowe, 'Postglacial environmental history of Rannoch Moor, Scotland, II: pollen diagrams and radiocarbon dates from Rannoch Station and Corrow areas', *Journal of Biogeography*, 6 (1979), 349–62.

13 H. Godwin, 'History of the natural forests of Britain: establishment, dominance and destruction', *Philosophical Transactions of the Royal Society of London*, B (1975), 47–67; G. F. Peterken, *Woodland Conservation and Management*, London, 1981; O. Rackham, *Trees and Woods in the British Landscape*, London, 1976; O. Rackham, *The History of the Countryside*, London, 1986.

14 R. Tittensor, 'History of Loch Lomond oakwoods', *Scottish Forestry*, 24 (1970), 100–18; J. M. Lindsay, 'The history of oak coppice in Scotland', *Scottish Forestry*, 29 (1975), 87–93.

15 B. Huntley and H. J. B. Birks, 'The past and present vegetation of the Morrone Birkwoods NNR, Scotland', *Scottish Forestry*, 29 (1979), 87–93.

16 C. Caufield, 'Anger as peer fells ancient pine forest', *New Scientist*, 8 March 1984.

17 D. N. McVean, 'The ecology of Scots Pine in the Scottish Highlands', *Journal of Ecology*, 51 (1963), 671–86; G. F. Peterken, *Woodland Conservation and Management*, London, 1981.

18 P. D. Moore and D. J. Bellamy, *Peatlands*, London, 1974; P. D. Moore (ed.), *European Mires*, London, 1984.
19 Nature Conservancy Council, *Nature Conservation in Great Britain*, Peterborough, 1986.
20 A. R. Clapham (ed.), *Upper Teesdale: The Area and its Natural History*, London, 1978.
21 D. A. Ratcliffe, *A Nature Conservation Review*, Cambridge, 1977.
22 N. Picozzi, 'Grouse bags in relation to the management and geology of heather moors', *Journal of Applied Ecology*, 5 (1968), 483–8; G. R. Miller and A. Watson, 'Heather productivity and its relevance to the regulation of Red Grouse populations', in Heal and Perkins (eds), *Production Ecology of British Moors and Montane Grasslands*,
23 P. Anderson and D. W. Yalden, 'Increased sheep numbers and the loss of heather moorland in the Peak District, England', *Biological Conservation*, 20 (1981), 195–214.
24 M. Rawes, 'Changes in two high altitude blanket bogs after cessation of sheep grazing', *Journal of Ecology*, 71 (1983), 219–35.
25 T. T. Elkington, 'Effects of excluding grazing animals from grassland on sugar limestone in Teesdale, England', *Biological Conservation*, 20 (1981), 25–36.
26 D. F. Ball, J. Dale, S. Sleail and O. W. Heal, *Vegetation Change in Upland Landscapes*, Bangor, 1982.
27 G. Sinclair (ed.), *The Upland Landscapes Study*, Environment Information Services, Dyfed, for the Countryside Commission, 1983.
28 A. Woods, *Upland Landscape Change: A Review of Statistics*, Cheltenham, 1984.
29 M. L. Parry (ed.), *Surveys of Moorland and Roughland Change*, (13 volumes), University of Birmingham, 1982.
30 G. Searle, 'Copper in Snowdonia National Park', in P. J. Smith (ed.), *The Politics of Physical Resources*, Harmondsworth, 1975.
31 'Conflict of interests: concern as military land hunger grows', *Countryside Commission News*, 21 (1986), 1.
32 M. Blacksell and A. Gilg, *The Countryside: Planning and Change*, London, 1981.
33 M. MacEwen and G. Sinclair, *New Life for the Hills*, London, 1983.
34 D. A. Stroud, T. M. Reed, M. W. Pienkowski and R. A. Lindsay, *Birds, Bogs and Forestry: The Peatlands of Caithness and Sutherland*, Peterborough, 1987; Nature Conservancy Council, *The Flow Country: The Peatlands of Caithness and Sutherland*, Peterborough, 1988.
35 M. Marquiss, D. A. Ratcliffe and R. Roxburgh, 'The numbers, breeding success and diet of Golden Eagles in southern Scotland', *Biological Conservation*, 34 (1985), 121–40.
36 C. T. Bibby, 'Merlins in Wales: site occupancy and breeding in relation to vegetation', *Journal of Applied Ecology*, 23 (1986), 1–12.
37 Nature Conservancy Council, *Nature Conservation and Afforestation in Great Britain*, Peterborough, 1986.
38 R. H. Grove, *The Future for Forestry*, Cambridge, 1984; P. J. Moore, 'The real world of private forestry', *ECOS: A Review of Conservation*, 6 no. 2 (1985), 2–7.
39 P. J. Moore, 'A victory for whom?', *ECOS: A Review of Conservation*, 9 no. 2 (1988), 1.

Chapter 2

This chapter draws heavily on material compiled by our colleagues in the Chief Scientist's Directorate of the Nature Conservancy Council. We would like to thank Hilary Allison, Lynne Farrell, John Hopkins, Richard Lindsay, Chris Newbold and John Robinson. Bob Evans and Michael Luscombe of the Department of Geography, Cambridge University helped sort out questions relating to soils.

Chapter 3

1 J. B. Sissons, R. A. Cullingford and D. E. Smith, 'Late-glacial and post-glacial shorelines in southeast Scotland', *Transactions of the Institute of British Geographers*, 39 (1966) 9–18; J. J. Donner, 'Land/sea level changes in Scotland', in D. Walker and R. West (eds), *Vegetational History of the British Isles*, Cambridge, 1970; J. M. Gray, 'Late-glacial and post-glacial shorelines in western Scotland', *Boreas*, 3 (1974), 129–38; J. B. Sissons, 'Shorelines and isostasy in Scotland', in D. E. Smith and A. G. Dawson (eds), *Shorelines and Isostasy*, London, 1983.
2 G. Cambers, 'Temporal scales in coastal erosion systems', *Transactions of the Institute of British Geographers*, new series 1 (1976), 246–58; A. H. W. Robinson, 'Erosion and accretion along part of the Suffolk Coast of East Anglia', *Marine Geology*, 37 (1980), 133–46; J. A. Steers *et al.*, 'The storm surge of 11 January 1978 on the east coast of England', *Geographical Journal*, 145 (1979), 192–205; I. Shennan, 'Flandrian and Late Devensian sea-level changes and crustal movements in England and Wales', in Smith and Dawson (eds), *Shorelines and Isostasy*.
3 C. W. Phillips (ed.), *The Fenland in Roman Times*, London, 1970; B. Simmons, 'Iron Age and Roman coasts around the Wash', in F. H. Thompson (ed.), *Archaeology and Coastal Change*, London, 1980; R. S. Seale, *Soils of the Ely District*, Harpenden, 1975.
4 C. F. Everard, 'On sea level changes', in Thompson (ed.), *Archaeology and Coastal Change*. M. Barth and J. Titus (eds), *Greenhouse Effect and Sea Level Rise*, New York, 1984; National Academy of Science, *Glaciers, Ice Sheets and Sea Level: Effect of a CO_2 Induced Climatic Change*, Washington D.C., 1985; M. F. Meier, 'Contribution of small glaciers to global sea level', *Science*, 226 (1984), 1418–21.
5 R. W. Horner, 'Current proposals for the Thames barrier and the organisation of the investigation', *Philosophical Transactions of the Royal Society of London*, A 272 (1972), 179–85; M. Pollard, *North Sea Surge: The Story of the East Coast Floods of 1953*, Suffolk, 1978.
6 M. A. Arber, 'Landslips near Lyme Regis', *Proceedings of the Geologists Association* 84 (1973), 121–27; D. Brunsden and D. K. C. Jones, 'The evolution of landslide slopes in Dorset', *Philosophical Transactions of the Royal Society*, A 253 (1976), 605–31.
7 E. M. Ward, 'The evolution of the Hastings coastline', *Geographical Journal*, 56 (1920), 107–23; E. R. Mathews, *Coast Erosion and Protection*, 3rd edn, London, 1934; K. Clayton, 'Salvation from the Sea', *Geographical Magazine*, 10 (1977), 622–25.
8 J. A. Steers, *The Coastline of Scotland*, Cambridge, 1973.
9 R. W. Horner, 'The Thames Barrier Project', *Geographical Journal*, 145 (1979), 242–53; J. P. Jolliffe, 'Coastal erosion and flooding: what are the broad options?', *Geographical Journal*, 149 (1979), 62–7.

10 G. Fowler, 'The extinct waterways of the Fens', *Geographical Journal*, 83 (1938), 30–9; H. Godwin, *Fenland: Its Ancient Past and Uncertain Future*, Cambridge, 1978.
11 H. C. Darby, 'The human geography of the Fenland before drainage', *Geographical Journal*, 80 (1932), 420–35.
12 T. Gray, *Buried City of Kenfig*, 1909; L. S. Higgins, 'An investigation of the sand dunes of the South Wales coast', *Archaeologia Cambrensis*, 1933.
13 W. Ritchie, 'Where land meets the sea', *Geographical Magazine*, 14 (1981), 772–4; R. Dolan, 'Erosion hazards along the mid-Atlantic coast', in R. G. Craig and J. L. Craft (eds), *Applied Geomorphology*, London, 1982; J. Hecht, 'America in peril from the sea', *New Scientist*, 118 no. 1616 (1988), 54–9.
14 T. Sheppard, *Lost Towns of the Yorkshire Coast*, London, 1912; H. Valentin, 'Land loss at Holderness', in J. A. Steers (ed.), *Applied Coastal Geomorphology*, London, 1971; A. D. Smith, 'Canute . . . your time is near', *The Guardian*, June 23, p. 38.
15 J. A. Steers *et al.*, 'The storm surge of 11 January 1978 on the east coast of England', *Geographical Journal*, 145 (1979), 192–205.
16 J. A. Steers, *Coastal Features of England and Wales*, Cambridge, 1981.
17 V. J. May, 'The retreat of chalk cliffs', *Geographical Journal*, 137 (1971), 203–6.
18 E. W. Gilbert, *Brighton: Old Ocean's Bauble*, London, 1954; M. J. Clark, 'Geomorphology in coastal zone management', *Geography*, 63 (1978), 273–83.
19 J. A. Steers and A. T. Grove, 'Shoreline changes on the marshland coast of north Norfolk, 1951–53', *Transactions of the Norfolk and Norwich Naturalist Society*, 17 (1954), 322–6; J. A. Steers *et al.*, 'The storm surge of 11 January 1978'; H. M. Allison, 'The Holocene evolution of Scolt Head Island, Norfolk', Ph.D. thesis, Cambridge University, 1985.
20 T. O'Riordan, 'The Yare Barrier proposals', *Ecos*, 1 no. 2 (1980), 8–14.

Chapter 4

1 Countryside Commission, *Recreation 2000: Enjoying the Countryside. A Consultation Paper on Future Policies*, Cheltenham, 1987.
2 Countryside Commission, *Recreation 2000*.
3 For example, see: J. Blunden and N. Curry, *The Changing Countryside*, London, 1985; A. Rogers, J. Blunden and N. Curry, *The Countryside Handbook*, London, 1985; P. Lowe, G. Cox, M. MacEwan, T. O'Riordan and M. Winter, *Countryside Conflicts: The Politics of Farming, Forestry and Conservation*, Aldershot, 1986; M. Shoard, *The Theft of the Countryside*, London, 1980; W. M. Adams, *Nature's Place: Conservation Sites and Countryside Change*, London, 1986.
4 M. Shoard, *This Land is Our Land: The Struggle for Britain's Countryside*, London, 1987.
5 For a full account see: J. K. Walton and J. Walvin (eds), *Leisure in Britain 1780–1939*, Manchester, 1983.
6 Examples of good statistical sources covering the period are: K. K. Sillitoe, *Planning for Leisure*, London, 1969; Countryside Commission, *Digest of Countryside Recreation Statistics 1979*, Cheltenham, 1979; Countryside Commission, *National Countryside Recreation Survey 1984*, Cheltenham, 1985; Countryside Commission, *Access to the Countryside for Recreation and Sport*, Cheltenham, 1986.
7 For example, compare the following. (The first edition, though the tables are out of date, is a standard descriptive introduction to recreation in Britain and for a basic view remains superior to the second.) A. J. Patmore, *Land and Leisure*, Harmondsworth, 1972, and *Recreation and Resources: Leisure Patterns and Leisure Places*, Oxford, 1983.
8 Countryside Commission, *Recreation 2000*.
9 Countryside Commission, *Recreation 2000*.
10 Countryside Commission, *Recreation 2000*.
11 J. Dower, *National Parks in England and Wales*, London, 1945; A. Hobhouse, *National Parks (England and Wales): Report of the Committee*, London, 1947.
12 The 1968 Act required all ministers 'to have regard to the desirability of conserving the natural beauty and amenity of the countryside'.
13 See note 6.
14 See note 3.
15 Shoard, *This Land is Our Land*.
16 P. L. Owens, 'Conflict as a social interaction process in environment and behaviour research: the example of leisure and recreation research', *Journal of Environmental Psychology*, 5 (1985), 243–59.
17 P. L. Owens, 'Recreational conflict and the behaviour of coarse anglers and boat users in the Norfolk Broads', Ph.D. thesis, Norwich: University of East Anglia, 1983.

Chapter 5

1 P. J. Bull, 'The spatial components of intra-urban manufacturing change: suburbanization in Clydeside, 1958–1968', *Transactions of the Institute of British Geographers* new series 3 (1978), 91–100.
2 P. E. Lloyd and C. M. Mason, 'Manufacturing industy in the inner city: a case study of Greater Manchester', *Transactions of the Institute of British Geographers*, new series 3 (1978), 66–90.
3 Lloyd and Mason, 'Manufacturing industry in the inner city'.
4 Paul Crawford, Stephen Fothergill and Sarah Monk, *The Effect of Business Rates on the Location of Employment: Final Report*, Cambridge, 1985.
5 S. Fothergill, M. Kitson and S. Monk, *Urban Industrial Change: The Causes of the Urban-Rural Contrasts in Manufacturing Employment Trends*, London, 1985.
6 D. Massey, *Spatial Divisions of Labour: Social Structures and the Geography of Production*, London, 1984.
7 D. W. Owen, M. G. Coombes and A. E. Gillespie, 'The differential performance of urban and rural areas in the recession' *Discussion Paper No 49 Centre for Urban and Regional Development Studies* Newcastle-upon-Tyne, 1983.
8 M. J. Breheny and R. W. McQuaid 'The M4 corridor: patterns and causes of growth in high technology industries' *Geographical Papers No 47*, Reading University, 1985.
9 Department of Trade and Industry, *Regional Industrial Policy: Some Economic Issues*, London, 1983.
10 R. P. Oakey, A. T. Thwaites and P. A. Nash, 'Technological change and regional development: some evidence on regional variations in product and process innovation', *Environment and Planning, A*, 14 (1982), 1073–86.
11 G. Gudgin, R. Crum and S. Bailey, 'White-collar employment in UK manufacturing industry', in P. W. Daniels (ed.), *Spatial Patterns of Office Growth and Employment*, Chichester, 1979.
12 J. R. L. Howells, 'The location of research and development: some observations and evidence from Britain', *Regional Studies*, 18 (1984), 17.

13 I. J. Smith, 'The effect of external takeovers on manufacturing employment change in the northern region between 1963 and 1973', *Regional Studies*, 13 (1979), 421–37.
14 J. N. Marshall, 'Research policy and review 4. Services in a postindustrial economy', *Environment and Planning, A*, 17 (1985), 1155–67.
15 R. D. Macey, *Job Generation in British Manufacturing Industry: Employment Change by Size of Establishment and by Region*, London, 1982.
16 M. J. Healey and D. J. Clark, 'Industrial decline in a local economy: the case of Coventry, 1974–1982', *Environment and Planning, A*, 17 (1985), 1351–67.
17 R. J. Pounce, *Industrial Movement in the United Kingdom, 1966–75*, London, 1981.
18 T. Killick, 'Manufacturing plant openings 1976–80', *British Business*, 17 June 1983, 466–8.
19 A. C. McKinnon, 'The development of warehousing in England', *Geoforum*, 14 (1983), 389–99.

Chapter 7

1 Department of Energy, *Energy Policy: A Consultative Document*, London, 1978.
2 House of Commons Select Committee on Energy, The Coal Industry, First Report Session 1986/87, London, 1987.
3 G. Leach, C. Lewis, F. Romig, A. van Buren and R. Foley, *A Low Energy Strategy for the United Kingdom*, London, 1979.
4 House of Commons Select Committee on Energy, The Energy Efficiency Office, London, 1985.
5 *Digest of UK Energy Statistics 1987* (HMSO).
6 G. Leach *et al.*, 1979; D. Pooley, 'Where will our energy come from?', paper presented at British Association for the Advancement of Science Annual Meeting, August 1985.
7 Power stations convert primary fuel (e.g., coal) at a maximum efficiency (limited by thermodynamics) of about 35 per cent. The rest is lost as 'waste' heat. But it is possible by reducing the efficiency of electricity generation to about 31 per cent to produce water at a temperature high enough for space heating of buildings if distributed to them in a pipe network. Production of electricity and useful heat in this way is known as combined heat and power generation (CHP). It is widely used in continental Europe but progress in Britain has been very slow.
8 *Digest of UK Energy Statistics 1985* (HMSO).
9 C. Robinson and J. Morgan, *North Sea Oil in the Future*, London, 1978.
10 O. Noreng, *The Oil Industry and Government Strategies in the North Sea*, London, 1980.
11 Shetland and Orkney, for example, had well established industries such as fishing and fish processing, knitwear and crofting. There is evidence that some indigenous industry suffered as a result of oil developments. See, for example, I. M. McNicoll, 'The pattern of oil impact on affected Scottish rural areas', *Geographical Journal*, 150 no. 2 (1984), 213–20, and H. D. Smith, A. Hogg and A. M. Hutcheson, 'Scotland and offshore oil: the developing impact', *Scottish Geographical Magazine*, 92 (1976), 75–91.
12 Department of Energy, *Coal for the Future: Progress with Plan for Coal and Prospects to the Year 2000* London, 1977.
13 Royal Commission on Environmental Pollution, Sixth Report, *Nuclear Power and the Environment*, London, 1976.
14 Sir Frank Layfield, *Sizewell 'B' Public Inquiry Report: Summary and Recommendations*, London, 1986.
15 British Coal Corporation (BC), *New Strategy for Coal*, London, 1986.
16 British Coal Corporation, *Annual Report and Accounts, 1988.*
17 For 'free market views' see K. Boyfield, *Put Pits into Profit: Alternative Plan for Coal*, London, 1985; C. Robinson and E. Marshall, *What Future for British Coal?*, London, 1981; R. Steenblik, 'The British coal industry and the wide world', paper presented at a conference *Coal on the Energy Seesaw*, April 1987. London, UK Centre for Economic and Environmental Development. For more interventionist arguments, see A. Glynn, 'The economic case against pit closures', NUM, Sheffield, 1985; S. Fothergill, 'Alternative job prospects', paper presented at a conference *Economic Prospects for the Coal Fields*, Sheffield, November 1985. London, Town and Country Planning Association.
18 House of Commons Select Committee on Energy, *The Coal Industry*, London, 1986/7.
19 'Risk' is a combination of the probability of an event and its consequences. Risk perception studies show that not only do people distrust experts' assessments of the former, but they are very concerned when consequences are potentially 'catastrophic', however minute the estimated probability (thus nuclear accidents are more feared than road accidents). Estimation of probability itself must rely on some subjective judgement.
20 'If it can happen it will!' – for example, it is impossible fully to incorporate the 'human factor'. Both the Three Mile Island and Chernobyl accidents resulted from a combination of human error and shortcomings in equipment.
21 M. Flood and R. Grove White, *Nuclear Prospects: a Comment on the Individual, the State and Nuclear Power*, London, 1976.
22 J. P. Stern, *UK Energy Issues 1987–92*, 1987.
23 Energy Technology Support Unit (ETSU), *Prospects for the Exploitation of the Renewable Energy Technologies in the United Kingdom*, AERE Harwell, Oxfordshire, 1985.
24 Nuclear power was not initially economically competitive with established means of producing electricity, but the investment was considered worthwhile because it was anticipated that costs would fall. The Central Electricity Authority (predecessor to the CEGB) had to be persuaded by the Government to construct Magnox rather than coal fired power stations in the mid-1950s.
25 J. K. Wright and R. H. Taylor, 'Electricity generation options from alternative sources', *Atom*, 339 (1985), 2–7.

Chapter 8

1 M. J. Liddle, 'An approach to objective collection and analysis of data for comparison of landscape character', *Regional Studies*, 10 (1976), 173–81.
2 The development of landscape evaluation since 1930 is reviewed in D. G. Robinson *et al*, *Landscape Evaluation: The Landscape Evaluation Research Project 1970–1975*, Centre for Urban and Regional Research, University of Manchester, 1976. Other useful discussions of various methods are given by R. S. Crofts and R. U. Cooke, *Landscape Evaluation: A Comparison of Techniques*, Department of Geography, University College London, 1974, and by E. C. Penning-Rowsell, *Alternative Approaches to Landscape Appraisal and Evaluation*, Middlesex Polytechnic Planning Research Group Report No. 11, Enfield, 1973.
3 One example is a survey originally planned as a land utilisation study but later used to produce a 'Wildscape

Atlas' – see A. Coleman, 'The conservation of wildscape: a quest for facts', *Geographical Journal*, 136 (1970), 199–205.

4 D. I. Linton, 'The assessment of scenery as a natural resource', *Scottish Geographical Magazine*, 84 (1968), 219–238.

5 A. W. Gilg, 'A critique of Linton's method of assessing scenery as a natural resource', *Scottish Geographical Magazine*, 90 (1974), 125–9, and 'The objectivity of Linton type methods of assessing scenery as a natural resource', *Regional Studies*, 9 (1975), 181–91.

6 B. A. Duffield and M. L. Owen, *Leisure + Countryside = : A Geographical Appraisal of Countryside Recreation in Lanarkshire*, Department of Geography, University of Edinburgh, 1970 and, M. L. Owen *et al.*, *TRRIP Series No. 1 System Description*, Tourism and Recreation Unit, University of Edinburgh, 1973.

7 For example, there was a comparison of its usefulness in a contrasting region which concluded it could have wider applications if modified: M. Blacksell and A. W. Gilg, 'Landscape evaluation in practice – the case of south-east Devon', *Transactions of the Institute of British Geographers*, 66 (1975), 135–40.

8 C. Tandy, *Landscape Evaluation Technique*, Croydon, 1971.

9 B. W. Avery *et al.*, *Soil Map of England and Wales, 1:1,000,000*, Ordnance Survey, Southampton, 1975.

10 Ministry of Agriculture, Fisheries and Food, *Agriculture Land Classification of England and Wales*, Agriculture Development Advisory Service, Land Service, MAFF, London, 1974.

11 J. S. Bibby, in M. F. Thomas and J. T. Coppock (eds), *Land Assessment in Scotland*, Aberdeen.

12 R. D. L. Toleman and D. G. Pyatt, in *Proceedings of the Tenth Commonwealth Forestry Conference*, Forestry Commission, London.

13 R. G. H. Bunce and R. S. Smith, *An Ecological Survey of Cumbria*, Kendal, 1978; R. G. H. Bunce *et al.*, 'The application of multivariate analysis to regional survey', *Journal of Environmental Management*, 3 (1975), 151–65.

14 R. G. H. Bunce *et al.*, 'An integrated system of land classification', *Annual Report of the Institute of Terrestrial Ecology 1980*, Abbots Ripton, 1981; M. O. Hill *et al.*, 'Indicator species analysis, a divisive polythetic method of classification and its application to a survey of native pinewoods in Scotland', *Journal of Ecology*, 63 (1975), 597–613.

15 R. G. H. Bunce *et al.*, *Land Classes in Great Britain: Preliminary Description for Users of Land Classification*, Merlewood Research and Development Paper No. 86, Grange-over-Sands, 1981.

16 C. B. Benefield and R. G. H. Bunce, *A Preliminary Visual Presentation of Land Classes in Britain*, Merlewood Research and Development Paper No. 91, Grange-over-Sands, 1982.

17 Alternatively, the procedure described by R. G. H. Bunce and R. S. Smith, *An Ecological Survey of Cumbria*, can be used to classify individual squares, using the key attributes identified by the analysis (Barr, in press).

18 C. P. Mitchell *et al.*, 'Land availability for production of wood energy in Great Britain', in A. Strub *et al.* (eds), *Energy from Biomass*, Applied Science Press, London, 1983.

19 Summarised by C. P. Mitchell *et al.*, 'Land availability'.

20 R. G. H. Bunce *et al.*, 'An ecological classification of land – its application to planning in the Highland Region, Scotland', in F. T. Last (ed.), *Land and Its Changing Uses: An Ecological Appraisal*, London, 1986.

21 R. G. H. Bunce *et al.*, 'Models for predicting changes in rural land use in Great Britain', in D. Jenkins (ed.), *Agriculture and the Environment*, Abbots Ripton, 1985.

22 R. Best, 'The extent and growth of urban land', *Planner*, 1 (1976), 8–11.

23 Ministry of Agriculture, Fisheries and Food, *Annual Review of Agriculture 1983* (Cmnd 8804), HMSO, London, 1983; Ministry of Agriculture, Fisheries and Food, *Agricultural Statistics, United Kingdom 1982*, HMSO, London, 1983.

Photographs

The photographs in this volume are drawn from the collection of the Cambridge University Committee for Aerial Photography, except where noted differently below. Photographs from the C.U.C.A.P. collection are in the copyright of the University of Cambridge, with the exception of those noted as 'Crown ©', indicating Crown Copyright/RAF photos reproduced by permission of the Controller of Her Majesty's Stationery Office. The C.U.C.A.P. reference code is given for all relevant photographs below.

Introduction

p.9 Pershore RAF Station, Worcestershire
(Courtesy of the Trustees of the Imperial War Museum)

p.10 Bolsover Castle, Derbyshire
(Reproduced by permission of the Syndics of Cambridge University Library)

1 Ben Nevis and Fort William. BUD 36 July 1975
2 Loch Avon and the Cairngorm Plateau. AGP 34 July 1972 (Crown ©)
3 Hirnant in the Berwyn in North Wales. AET 51 May 1962
4 Beinn Eighe and Liathach in Wester Ross. BSP 44 May 1975
5 Derry Lodge, Deesside, Cairngorms. BOW 11 Oct 1973
6 Rannoch Moor. AAK 17 July 1979
7 Craigellachie National Nature Reserve. AHZ 5 July 1963
8 Loch Rannoch and the Black Wood of Rannoch. BRI 4 Aug 1974
9 Cow Green Reservoir in Upper Teesside. BKX 42 Oct 1972
10 Beinn A'Ghlo. CAD 1 July 1976
11 Managed grouse moorland, north of Balmoral. AGP 41 July 1962 (Crown ©)
12 Oxendale, Langdale Valley in the Lake District. BYG 31 June 1976
13 Ski lifts and runs on Cairngorm. BGY 7 July 1971
14 High Cup Nick in the North Pennines. BME 45 April 1973
15 Lynmouth on the North Devon coast. ACU 67 July 1976
16 Silver Flowe in Kirkudbrightshire. AKI 11 July 1964
17 Afforestation near Rochester in the Cheviots. ATH 13 July 1967
18 Llynn Brianne reservoir, north of Llandovery. BLA 93 Nov 1972
19 Peaks of the Brecon Beacons. BLA 81 Nov 1972
20 Baosbheinn in Torridon. BSQ 20 May 1975
21 Weeting Heath, Norfolk. WT 81 June 1958 (Crown ©)
22 Bradfield Woods, Suffolk. BMO 17, BMO 10 May 1973
23 Moccas Park, Hereford. BMX 74 June 1973
24 Thursley Common, Surrey. BSU 46 June 1975
25 Wyre Forest, Shropshire and Worcestershire. RC8-CF 190 July 1977
26 Wickham Skeith, Suffolk. BYI 54 June 1976
27 Madmarston Hill, Swalcliffe, Oxfordshire. ACT 71, ATZ 42 Apr 1961, 1968 (Crown ©)
28 Wimpole Hall and Park, Cambridgeshire. ER064 (Crown ©), CEI 67
29 Radcot, Oxfordshire. BVV 37 Aug 1975
30 Ford, Northumberland. BVM 100 Aug 1975
31 Ranworth Marshes, Woodbastwick, Norfolk. WJ 35 June 1958 (Crown ©)
32 Wedholme Flowe, Cumbria. RB 12 July 1955 (Crown ©)
33 Standlake and Stanton Harcourt, Oxfordshire. LW 49 June 1953 (Crown ©)
34 Site of Snelshall Priory, Buckinghamshire. ATZ 2 Apr 1968
35 Farmland south east of Blyth, Nottinghamshire. CCG 67 May 1977
36 Coastlines of emergence, Iona, Inner Hebrides. QX 58 July 1955 (Crown ©)
37 Retreating coastlines, Overstrand, Norfolk. CQ 07 June 1949 (Crown ©)
38 Methwold Fen, near Southery, Cambridgeshire. RC8-EB 87 Mar 1982
39 Oil refineries at Coryton. CPT 21 Oct 1982
40 Black Ven in Dorset. BIN 15 June 1972
41 Longshore drift, Hastings, Sussex. CER 7 Sept 1977
42 The Mull of Oa, Islay, Inner Hebrides. BOX 81 Oct 1973
43 Thames Barrier, looking towards Woolwich. CQY 28 Oct 1986
44 Fenland. AXI 49 Apr 1969
45 Kenfig Burrows, between Margam and Porthcawl, South Wales. BXD 51 June 1976
46 Holderness, looking south towards Tunstall. FV 2 June 1951 (Crown ©)
47 Scolt Head Island. CFA 44 Jan 1978
48 Orford Ness, Suffolk (1949). CQ 64 June 1949 (Crown ©)
49 Orford Ness, Suffolk (1983). RC8-EZ 82 Feb 1983
50 Birling Gap, Seven Sisters coast, Sussex. CNR 8 Oct 1980
51 Peacehaven, Sussex. BSV 21 June 1975
52 Wells-next-the-Sea, Norfolk. CFA 4 Jan 1978
53 Yare Barrier scheme, Breydon Water. CQ 43 June 1949 (Crown ©)
54 Woburn Abbey, Bedfordshire. AN 20 June 1948
55 Buxton, Derbyshire. OS 92 July 1954 (Crown ©)
56 Scarborough, North Yorkshire. BNG 71 June 1973
57 Cardiff, Cathays Park, South Glamorgan. AJA 63 June 1964
58 Grasmere, Cumbria. BYG 8 June 1976
59 Barton Broad, Norfolk. CGF 35 July 1978
60 Stonehenge, Wiltshire. CEH 48 Aug 1977
61 Snowdon, Gwynedd. SM 23 May 1956 (Crown ©)
62 Twyni Bach. CGN 3 July 1978
63 Oxwich, West Glamorgan. RC8-N 209 Oct 1969
64 Pennine Way, Greater Manchester. BSI 21 Apr 1975
65 Tissington Trail, Derbyshire. CRA 29 Dec 1986
66 Derwent Reservoirs, Derbyshire. CCF 82 May 1977
67 Trawsfynydd nuclear power station, Gwynedd. AKW 82 May 1965
68 Tunstead, Great Rocks Dale, Derbyshire. CRA 34 Dec 1986
69 Worbarrow Bay, Dorset. R36 June 1956 (Crown ©)
70 Lower Don Valley, Sheffield, South Yorkshire. RC8-DP 132 Sept 1980
71 Beeston, Nottinghamshire. AFO 52 June 1962 (Crown ©).
72 Halifax, West Yorkshire. BWD 68 Nov 1975
73 Fakenham, Norfolk. RC8-FI 131 June 1983

74 Dudley, West Midlands. COK 28
75 Cambridge, Cambridgeshire. CQA 40 May 1984
76 Gateshead, Tyne and Wear. CQS 16 June 1986
77 Invergordon, Highland. BOQ 7
78 Speke, Merseyside. COM 85 July 1981
79 Consett, Durham. BLK 70 Jan 1973
80 Hursley Park, Hampshire. RC8 AY 147 May 1975
81 Christchurch, Dorset (photograph kindly supplied by Plessey Defence Systems).
82 Bridgend, mid-Glamorgan. CQT 33 June 1986
83 Magor, Gwent. CQT 5 June 1986
84 Daventry, Northamptonshire. BOE 41 July 1973
85 Heathrow Airport, London. RC8 DZ 94 Oct 1981
86 Gravelly Hill Interchange, Birmingham. CHW 25 July 1976
87 Waverley Station, Edinburgh. CDD 5 July 1977
88 Crewe Yards, Cheshire. BFQ 10 June 1971
89 Devizes Locks, Wiltshire. CH 35 1949 (Crown ©)
90 Devizes Locks, Wiltshire. CQZ 89 1986
91 The Isle of Dogs, London. BIJ 76 May 1972
92 The Isle of Dogs, London. CQY 23 1986
93 Dover Eastern Docks, Kent. BJL 14 1972
94 Dover Eastern Docks, Kent. CQZ 7 1986
95 Ribbleshead Viaduct, Yorkshire. BLZ 59 1973
96 M25, London. CPP 20 1981
97 M25, London. CQZ 1 1986
98 Forth Road Bridge, Lothian. CKI 55 July 1979
99 Channel Tunnel: The Cheriton Terminal Site, Kent. CQZ 35 1986
100 Aerial thermographic image of part of Cardiff, South Wales (reproduced by kind permission of Clyde Surveys Limited, Maidenhead and Welsh Energy Efficiency Office, Cardiff).
101 Brent Oil and Gas Field, North Sea (photograph kindly supplied by Shell).
102 Steel platform construction, Nigg Bay, East Ross. BRD 77 Aug 1974
103 On-shore terminal, St Fergus, Aberdeenshire. BVE 55 July 1975
104 Wytch Farm, Purbeck Peninsula, Dorset. CQR 14 July 1985
105 Sizewell, Suffolk Coast. AIZ 62 June 1964
106 Selby Coalfield, Yorkshire. CQW 46 July 1986
107 River Trent Power Stations, Nottinghamshire. CQW 41 July 1986
108 Opencast coal mine, Hauxley, Northumberland. BQA 86 June 1974
109 Merthyr Tydfil, South Wales. CQU 10 June 1986
110 Belvoir Castle and the Vale of Belvoir, Leicestershire. CRA 2 Dec 1986
111 Sellafield nuclear reprocessing plant, Cumbria. CLB 2 Sept 1979
112 Experimental fast reactor, Dounreay, Caithness (photograph kindly supplied by the United Kingdom Atomic Energy Authority).
113 Pump storage scheme, Dinorwic, North Wales. RC8-GQ 31, RC8-GQ 24 June 1984
114 Wind generators, Carmarthen Bay, South Wales. (photograph kindly supplied by the CEGB)
115 Widdington, Northumberland. CPL 24 May 1982
116 Oxfordshire countryside near Witney. CML 61 July 1980
117 Severn Estuary. BNH 78 June 1973
118 Land Classes 1–2. Near Hawkesbury, Glos. AJL 10 July 1964
119 Land Classes 3–4. Near Market Deeping, Lincs. ABP 59 June 1960
120 Land Classes 5–6. Near Ashburton, Devon. FL 40 Aug 1950
121 Land Classes 7–8. Kingsbridge Estuary, Devon. BSI 105 Apr 1975
122 Land Classes 9–10. Near Stangton, Leics. DB 22 July 1949
123 Land Classes 11–12. Wisbech St Mary, Cambs. ATR 90 Aug 1967
124 Land Classes 13–14. Gravel pits, Great Linford, Bucks. BFQ 87 July 1971
125 Land Classes 15–16. Near Newtown, Powys. AYT 66 July 1969
126 Land Classes 17–18. Near Llanrug, Gwynedd. AUN 37 June 1968
127 Land Classes 19–20. Near Farndale, N. Yorks. AWG 66 Nov 1968
128 Land Classes 21–22. Near Rogart, Highalnd, over Strath Brora. BVJ 108 July 1975
129 Land Classes 23–24. Ben Nevis. BSQ 81 May 1975
130 Land Classes 25–26. Near Falkland, Fife. CHQ 6 Aug 1978
131 Land Classes 27–28. Near Cadder, Strathclyde. BKD 69 July 1972
132 Land Classes 29–30. Near Drumbeg, Highland. EVJ 73 July 1975
133 Land Classes 31–32. Fetlar, Shetland. AME 68 Aug 1965

Notes on Contributors

W.M. Adams is a Lecturer in Geography at the University of Cambridge. He obtained an MSc in conservation at University College London, and completed his PhD at Cambridge on the environmental and socio-economic impacts of water resource development in Nigeria. He also carried out research on conservation policy in Britain, and is author of *Nature's Place: Conservation Sites and Countryside Change*. He is currently working on the development of nature conservation policies in the wider countryside, and the way in which ideas of sustainable development influence protected area policy, both in Britain and abroad.

David Banister is Senior Lecturer in Transport Policy at University College London. He graduated from Nottingham University and completed his PhD at Leeds University in transport studies. He has published widely, including volumes on *Urban Transport and Planning* and *Transport in a Free Market Economy*. At present he is carrying out research into European competition policy, transport planning methods and evaluation.

Tim Bayliss-Smith is a Lecturer in Geography at the University of Cambridge. Since completing his PhD on the Solomon Islands, he has divided his research effort between the study of island ecology and subsistence agriculture in the Pacific and coastal studies in Britain. He is the author of *The Ecology of Agricultural Systems* and co-author of *Islands, Islanders and the World*.

Bob Bunce qualified in botany at University College of North Wales and has been employed at the Institute of Terrestrial Ecology's Merlewood Research Station, initially studying woodland vegetation. He has developed a computerised system for classifying Britain into 32 ecological classes, used for assessing the composition of the countryside, and for predicting actual and potential change.

Francine Hughes is a geographer. She completed her PhD at the University of Cambridge and has published research on the ecology of floodpain forests in Kenya and on the downstream impacts of dams. She has also worked for the Nature Conservancy Council on habitat loss in Britain. She is currently a Fellow of Newnham College, Cambridge and is continuing research on the ecology of floodplains.

Peter L. Owens is a Lecturer in Geography at the University of Sheffield. He graduated from Queen Mary College, London and completed his PhD in environmental sciences at the University of East Anglia. He has a major research interest in the theory of social and psychological conflict and its application in the management and resolution of rural resource conflicts. He also has a specialist knowledge of attitude and other social surveys and has published research on the problem of regional disparity in the European Community.

Susan Owens is a Lecturer in Geography at the University of Cambridge and a Fellow of Newnham College. She graduated in environmental sciences at the University of East Anglia and completed a PhD there on energy and land-use planning. Before taking up her current post, she carried out research for the Energy Panel of the Social Science Research Council and at the Institute of Planning Studies, University of Nottingham. She has published a book and numerous articles on energy and environmental issues and has recently undertaken a study of European Community energy and environmental policies for the European Commission.

George Peterken currently holds a Bullard Fellowship at Harvard University, but is ordinarily employed at the Nature Conservancy Council. For the last 19 years, he has dealt with policy, management and research issues relating mainly to woodland conservation and forestry. His publications have concentrated on British woodland history, management, classification and floristics, and include *Woodland Conservation and Management*. He holds a PhD and DSc from the University of London.

H.D. Watts is Senior Lecturer in Geography at the University of Sheffield. He is a graduate of the University of Leicester and completed his PhD at the University of Hull. He has published a number of books and papers on industrial change in Britain focussing upon the activities of large multi-national firms. He is currently engaged in studies of the geography of plant closures by large firms and the implications for local labour markets of the adoption of the new technologies.

Index

Photographs are indicated by page numbers in italics